Glyphosate 2023–2033

Norbert Fraeyman

Glyphosate 2023–2033

Springer

Norbert Fraeyman
Ghent University Hospital
Ghent University
Ghent, Belgium

ISBN 978-3-031-97539-4 ISBN 978-3-031-97540-0 (eBook)
https://doi.org/10.1007/978-3-031-97540-0

This Springer imprint is published by the registered company Springer Nature Switzerland AG
The registered company address is: Gewerbestrasse 11, 6330 Cham, Switzerland

To my wife: ***Janine***
To my children: ***Thomas and Sarah***
And their partners: ***Anneleen and Brecht***
To my grandchildren: ***Felix, Marie, Kamiel and Mona***

General Introduction

What is the point of writing a book on glyphosate after its use in the EU is granted for another 10 years until 2033?

From 2019 until 2023, The European Food Safety Authority (EFSA), The European Chemicals Agency (ECHA), and the Reviewing Member States scrutinized all the applicants' information with the underlying question of whether or not to prolong permission to use glyphosate. Many people not directly involved in this reviewing process, even the most interested, were overwhelmed by the multitude of European Regulations, Directives, and recommendations and in some way, lost their grip on the information. Science was divided and polarized into absolute pro or contra, and nuances were frequently "forgotten." In addition, the lay press was filled with heated debates among stakeholders, and positions unconditional in favor or against glyphosate were taken and vigorously defended, consuming much intellectual energy and money. All this together was not the most exalting spectacle to watch.

The obvious question is whether all this was needed. Was this the only way to proceed? If anything went wrong or if lessons have to be drawn from this period, we must make adjustments now. If not, we will face the same turmoil within less than 10 years. Now, in 2025, the dust of this storm has fallen, and it is the right time to evaluate the situation, not only from 2019 to 2023 but also from before.

The book is NOT intended to convince anyone of my point of view on the glyphosate story, although I will not refrain from communicating my opinion. The information analysis was performed objectively, and whenever I take a private stand, it will be clear to the reader. Is it NOT the aim to write a "nice" book! Some of the chapters are probably more difficult to master, but since the book's subject is largely based on the results of scientific research, this is inevitable. A second source of complexity is that glyphosate is a component of the entire chemical pollution of the environment. Both aspects are intertwined, which is reflected in the text and might hinder easy reading. Instead, the aim was to provide the readers with sufficient information allowing them to form their own personal view.

The first draft of this book was written from February to June 2023, i.e., during the months following the decision of the EFSA not to decide about glyphosate in

2022 but rather in December 2023 because of the number of documents supporting the demand. The trigger to finish the book was the publication of the advice of EFSA on 26/7/2023 in the EFSA journal allowing the further use of glyphosate until December 2033.

The book is organized into 11 chapters: basic information on glyphosate (Chap. 1), the relationship between exposure to glyphosate and human diseases (Chaps. 2, 3, 4, 5, and 6), the problem of multiple exposures (Chap. 7), and the relation to other aspects of the environment (Chap. 8). The advice of the EFSA and the supporting documents are critically examined in Chap. 9, and some opinions on how to proceed are given in Chap. 10. Chapter 11 contains information in addition to the data provided in the previous chapters. Chapters 2, 3, 8 and particularly Chap. 9 are more elaborate than the others because of their prime importance. Some chapters start with an introduction including basic textbook-like information on, e.g., carcinogenicity, the endocrine, or the neurological system. These introductions are not intended to be complete but offer a short acquaintance for less scientifically oriented readers. The interested reader will find references at will with full explanations.

The choice of references is a delicate matter since bias is possible. I took peer-reviewed recent references mainly from the last 4–6 years, together with some crucial older references. The validity of the information in websites and public communications is much more difficult to check; hence, these sites should always be approached cautiously. There is redundancy, as some landmark references are cited in more than one chapter. For high-level scientists, I have referred to recent references on each item. For the less high-level readers, I have tried to balance the intrinsic difficulty of each of the subjects and the readability. This has worked well for some chapters but probably not for all. For those offended because it is too difficult or too easy, I sincerely apologize.

Assuming the reader could agree that the evaluation of the past period is a sensible action and could contribute to the handling of future glyphosate dossiers, why should it be me to write this book?

Part of the reason is my earlier occupation within the University Hospital. Approximately 25 years ago, I was involved in the toxicology laboratory, monitoring patients admitted to the emergency department with acute intoxications. During that time, I was confronted with several suicide attempts, among others, with victims drinking Roundup, the commercial preparation containing glyphosate. Many of the victims survived the attempt but suffered from gastrointestinal discomfort, respiratory symptoms, hypotension, and headache; symptom monitoring and comfort treatment were frequently sufficient. The contrast between the limited acute toxicity of glyphosate compared to other pesticides, such as the organophosphate parathion or the carbamate aldicarb or paraquat was striking and intriguing. As a biochemist, I was also charmed by the approach of Monsanto, which not only produced glyphosate but also from 1996 on genetically modified vegetables resistant to glyphosate. At that time, genetic engineering techniques were cutting-edge technologies. From an economic point of view, this was a big step: killing unwanted herbs without damaging valuable crops in one treatment.

In 2015, the International Agency for Research on Cancer (IARC) shook the glyphosate community, saying that glyphosate was probably carcinogenic to humans. I studied for a long time neuroblastoma tumors in a mouse model, in cultured cells, and in children, and the possibility that glyphosate could cause cancer had far-fetched consequences not only in general but also in the context of the neuronal tumor in which I was interested. At that time, I was also involved in the work of the Superior Health Council in Belgium (SHC), which advises chancellors on matters of human health. The IARC statement prompted the SHC to take a stand on this matter, and I participated in writing the advice of the SHC. The advice was published in January 2020, i.e., at the time, the whole world was more interested in the all-emerging coronavirus crisis, and glyphosate was snowed in.

Although my point of view on glyphosate corroborates with the advice of the SHC, this book was written independently of this organization. It is my responsibility and does not necessarily reflects the vision of the organizations I am involved in today: the Superior Health Council, the Faculty of Medicine of the University of Ghent, and the University Hospital Ghent.

I am convinced that discussing a subject such as glyphosate has to be multidisciplinary, encompassing biochemistry, toxicology, genetics, medicine, and environmental issues, as well as agriculture, economics, and ethics. We have excellent books and top-level scientists bred and saddled in each of these topics, but I noticed that coherence among the different approaches is frequently lost. In contrast to these sophisticated texts on particular issues, this book addresses many subjects related to glyphosate, trying to offer a balanced, nuanced picture. I hope to remain friends with the specialists in each biomedical subject and become friends with the specialists in (for me) less obvious subjects.

The "What now?" question is relevant. First, if everything is left as it is now, we will face within 7 years (3 years before the end of the permitted period), a similar situation as we have had from 2020 to 2023. We need more evidence-based information to counteract the many controversial statements on the subject guided by economic motives; the earlier Monsanto papers are the undisputable evidence. Even in the highly esteemed literature, papers appeared that were written by authors in a direct relationship with Monsanto/Bayer. The fact that the Glyphosate Renewal Group has spent hundreds of thousands of euros in lobbying activities at least fuels the uneasy feeling of subjectivity. Second, the permit to use glyphosate for another 10 years has convinced some people that glyphosate is safe based on the authority of EFSA and consider any further activity unnecessary. However, the ongoing and often evidence based anxiety of many others, both groups or organisations and individuals is overruled; the search for (chemical) alternatives should not be considered as a second priority. New decisions will have to be made, and we need new and convincing evidence to do so.

I hope that this book can help in this respect.

Contents

Chapter 1
Basic Information on Glyphosate

Glyphosate, the Chemistry

Glyphosate is a simple organic molecule (for its structure, see Fig. 1.1a). It is composed of two moieties: glycine, an amino acid (Fig. 1.1b) coupled to a phosphate group (Fig. 1.1c shown as the ammonium salt of the ionic form). The synonym/IUPAC name is N-(phosphonomethyl)glycine, CAS No. 1071-83-6, EC No. 213-997-4, and Beilstein No. 2045054. The molecular weight is 169.07. The molecule has four ionizable functional groups with pKa1 = 2, pKa2 = 2.29, pKa3 = 5.96, and pKa4 = 10.98. The main form is acidic which occurs in different salt forms, isopropylamine, monoammonium, potassium, trimesium, dimethylammonium, or sesquisodium salt. Glyphosate should not be confused with the herbicide glyphosinate, which is structurally different.

Glyphosate was first synthesized in the 1950s, but its herbicidal properties were discovered in the 1970s. The original patent granted to Monsanto in 1974 on glyphosate as an herbicide expired in 2000. Since then, many generic glyphosate-containing products have been marketed by many companies.

Glyphosate is used as a formulation consisting of glyphosate in a salt form and an adjuvant, tallowamine being frequently used. The most renowned commercial formulation is Roundup. The role of the adjuvant is to facilitate the uptake of glyphosate, i.e., the herbicidal active compound, by the plant by modifying the cell membrane of the plants. If no adjuvant is used, glyphosate is washed away by rain without having a full effect on the plants.

An overview of the formulations, their names, and formulas is given in Fig. 1.2, and the overall schema of the application of glyphosate is shown in Fig. 1.3.

The number of commercially available formulations is high as shown in Table 1.1 (taken from Gill et al., 2018).

N. Fraeyman, *Glyphosate 2023–2033*,
https://doi.org/10.1007/978-3-031-97540-0_1

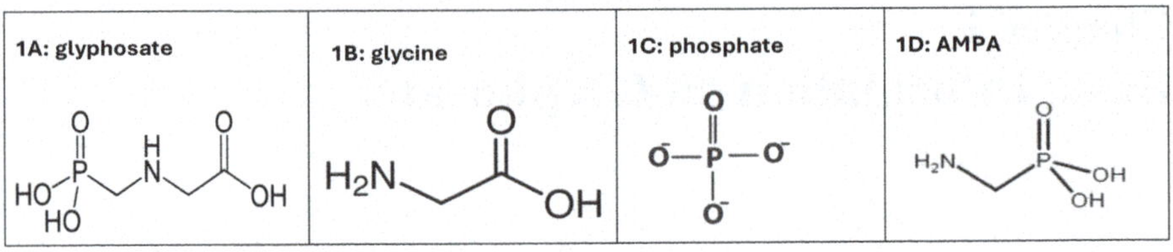

Fig. 1.1 Structural formulas of glyphosate and related chemicals

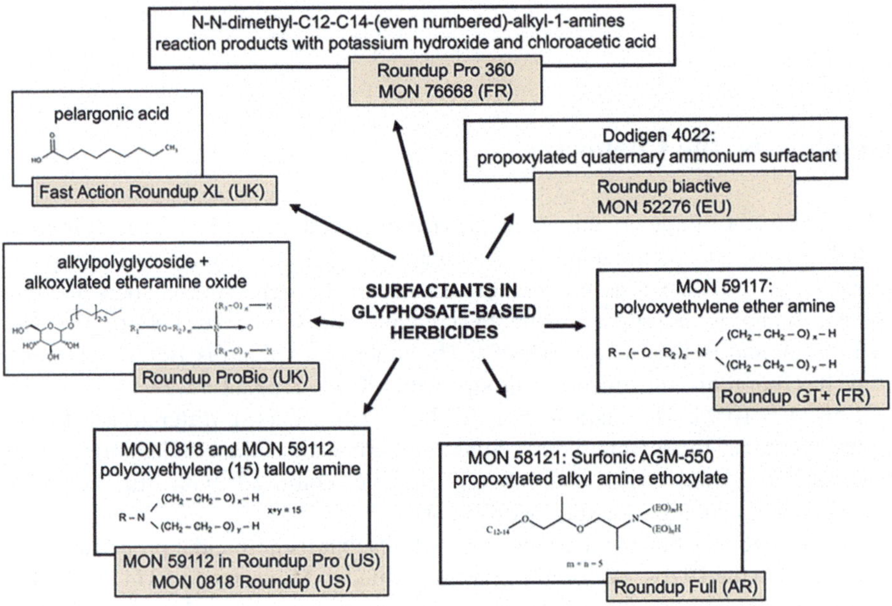

Fig. 1.2 Examples of formulations of glyphosate. (Taken from Mesnage et al. (2019))

The different salt forms of glyphosate determine the solubility of the compound in water, which is the most convenient application mode for the pesticide. When the salt is taken up by a plant, it is converted to an acidic form. Hence, the counterion for glyphosate acid does not affect its pesticidal activity, but different salt forms in combination with the adjuvant could significantly increase or decrease the rate of activity (Travlos et al., 2017). The action of the trimesium salt is reported to be very fast compared with, e.g., the sodium salt, which might be due to the higher solubility of the trimesium salt (approximately 1200 mg/L) compared with the solubility of the sodium salt (approximately 150 mg/L).

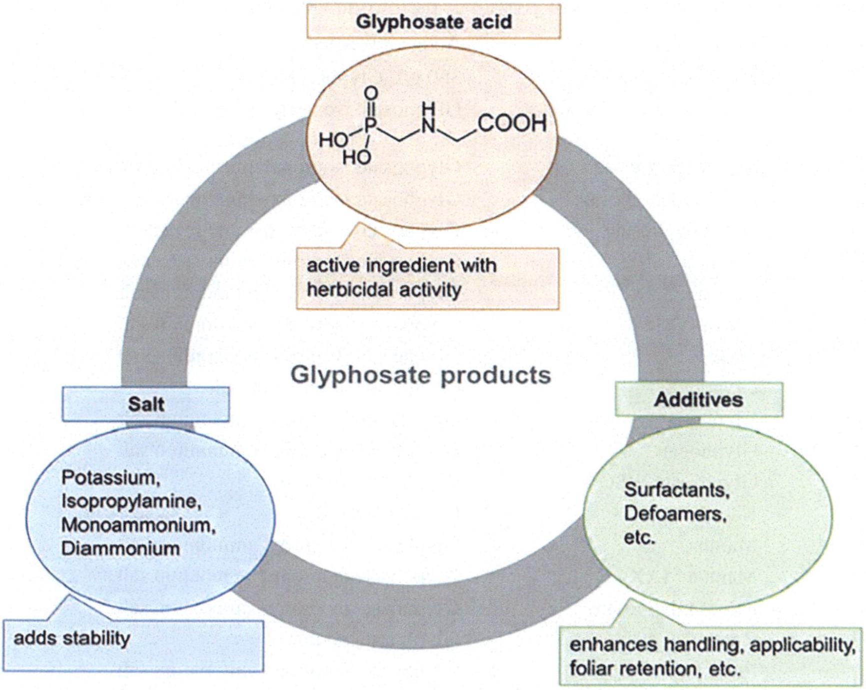

Fig. 1.3 Overall pattern of glyphosate use. (Taken from https://citrusindustry.net/2019/06/24/understanding-glyphosate-formulations/)

Glyphosate, the Biochemistry

When animal species, including humans, are exposed orally (food, drinks, etc.), glyphosate is absorbed by 30–36%; dermal absorption of glyphosate with an intact skin is only 1–3%, even at high doses. Glyphosate is poorly metabolized for approximately 1–2%. The majority of the ingested compound is secreted via the feces and the urine, and approximately 5% is retained in bone. The main metabolite is aminomethylphosphonic acid (AMPA, Fig. 1.1d), which is produced via the enzymatic activity of glyphosate oxidoreductase.

The biological activity of glyphosate is embedded in its reaction with a specific enzyme in a biochemical cycle called the shikimic acid pathway (see Fig. 1.4). This pathway is a seven-step succession of enzymatic reactions starting from phosphoenolpyruvate and erythrose-4-phosphate to chorismate and further to aromatic amino acids (L-tryptophan, L-phenylalanine, and L-tyrosine). This pathway is lost in animal cells, including humans, and aromatic amino acids have become essential components in the diet (Maeda & Dudareva, 2012). Glyphosate is a competitive inhibitor of one of the enzymes, i.e., 5-endopyruvylshilimate-3-phosphate synthase (EPSP synthase or EPSPS). EPSPS was also found to be present in evolutionarily

Table 1.1 Common formulations of glyphosate used worldwide (not exhaustive)

S. No	Brand name	Active ingredients
1.	Roundup® Renew	360 g/L Glyphosate
2.	Watson weed killer 360 concentrate	Glyphosate isopropyl amine salt (360 g/L)
3.	McGregor's Weed Out	Glyphosate isopropylamine salt (48%)
4.	AGPRO Glyphosate 360	Glyphosate isopropylammonium salt (360 g/L)
5.	Dow Glyphosate 360	Glyphosate isopropylamine salt (360 g/L)
6.	Clearout 180	Glyphosate isopropyl ammonium salt
7.	Clearout 360	Glyphosate isopropyl ammonium salt
8.	Cobra	Glyphosate isopropyl ammonium salt
9.	Erase	Glyphosate isopropyl ammonium salt
10.	Glygran SG	Glyphosate ammonium salt
11.	Glyphofix	Glyphosate isopropyl ammonium salt
12.	Glyphogan	Glyphosate isopropyl ammonium salt
13.	Glyphosate WSG	Glyphosate-sodium
14.	Kalach	Glyphosate-sodium
15.	Mamba	Glyphosate isopropyl ammonium salt
16.	Mamba MAX	Glyphosate isopropyl ammonium salt
17.	Nexus Glyphosate	Glyphosate isopropyl ammonium salt
18.	Rondo	Glyphosate-sodium
19.	Roundup	Glyphosate isopropyl ammonium salt
20.	Roundup Bio-dry	Glyphosate-sodium
21.	Roundup CT	Glyphosate isopropyl ammonium salt
22.	Roundup Turbo	Glyphosate isopropyl ammonium salt
23.	Roundup Ultra	Glyphosate isopropyl ammonium salt
24.	Slash Turbo	Glyphosate isopropyl ammonium salt
25.	Sting	Glyphosate isopropyl ammonium salt
26.	Touchdown	Glyphosate + trimesium
27.	Touchdown plus	Glyphosate + trimesium
28.	UAP Glyphosate	Glyphosate-sodium
29.	Back draft	Glyphosate + imazaquin
30.	Expert	Glyphosate + S-metolachlor + atrazine
31.	Extreme	Glyphosate + imazethapyr
32.	Flexstar GT	Glyphosate + formesafen
33.	Sequence	Glyphosate + S-metolachlor

lower cellular entities, such as microbes in the soil and the intestines of animals, including humans (Leino et al., 2021).

EPSPS can exist in multiple configurations due to genetic variability in the plants but also due to induced mutations as a reflex to exogenous environmental pressure leading to evolutionary adaptations. EPSPS in bacteria consists of two configurations: class I, which is sensitive to glyphosate, and class II, which is resistant to the pesticide. Within both classes, many isoforms are found due to altered genetic

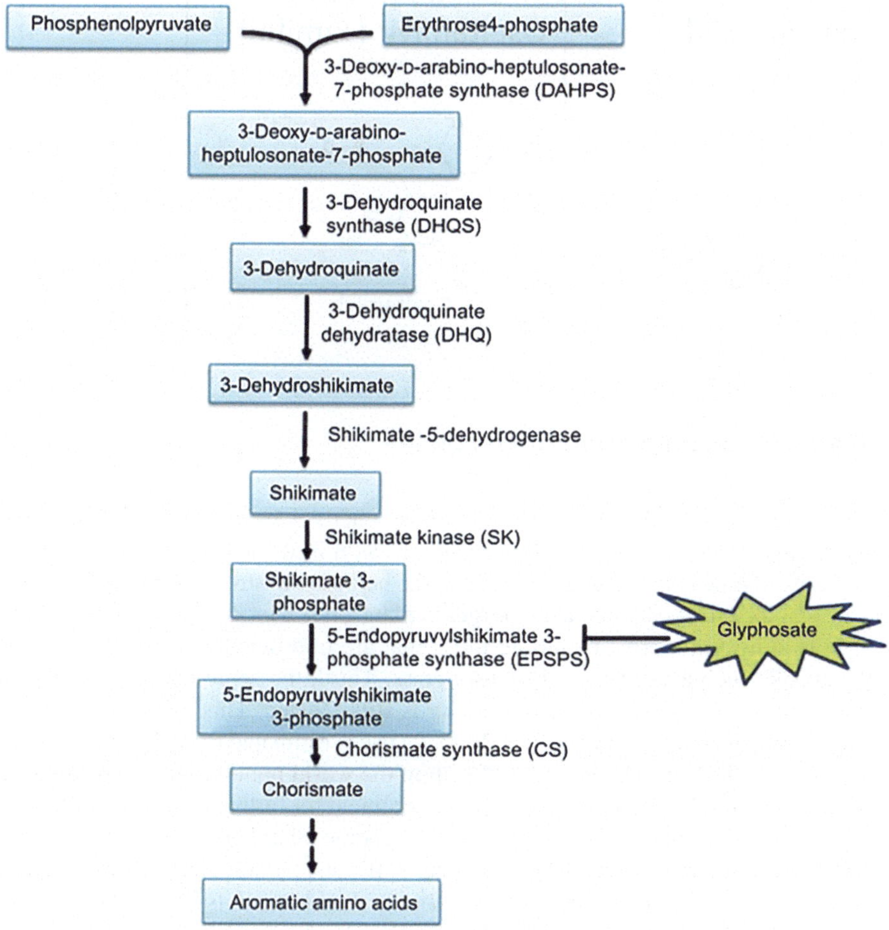

Fig. 1.4 Shikimic acid biochemical pathway. (Taken from Tiwari (2019) Cyanobacteria, from basic science to applications, book eds Mishra, Tiwari, and Rai. https://www.sciencedirect.com/topics/neuroscience/3-phosphoshikimate-1-carboxyvinyltransferase)

makeup. The existence of these two classes of bacteria in the microbiomes of rodents and honey bees including in the gut bacteria of humans makes them partly sensitive to glyphosate, which disturbs the critical equilibrium of bacteria in the (human) gut. The role of EPSPS molecular variability will be addressed in Chap. 6 of this book.

Shikimic acid or shikimate is, in its anionic form, an important plant metabolite and was first isolated from the Japanese flower shikimi-no-ki in 1885. Shikimic acid is a precursor for pharmaceutical compounds such as the antiviral drug oseltamivir (Tamiflu), which are used to combat bird flu infections.

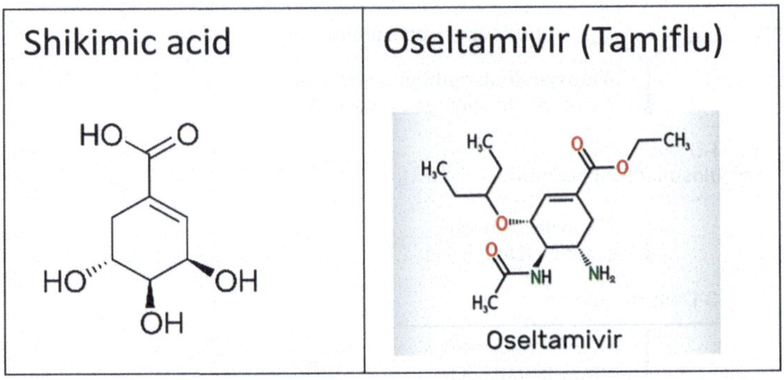

Glyphosate Is an Excellent Herbicide

Glyphosate is one of the best—if not the best—herbicides at all times mainly for three reasons (Duke & Powles, 2007, 2008). First, it is active throughout the whole plant, not only the leaves but also the roots; the herbs are destroyed entirely. Second, because of its universally occurring target in plants, it is a broad-spectrum herbicide that kills almost all herbs, including those that are traditionally resistant to herbicidal attacks such as kudzu, and johnsongrass. Third, the acute human toxicity of glyphosate is low, while long-term toxicity is still debated. This makes glyphosate the herbicide of choice since the year 2000. Its use is estimated to be 25% of all used herbicides worldwide. The need for "Feeding the world population" as sustainable development goal 2 is frequently used as an argument for further using the herbicide.

Glyphosate has taken this lead position from some other herbicides, such as atrazine (not used anymore in the EU), dicamba, 2.4D, and paraquat, mainly because of its low acute toxicity compared with the other herbicides and its herbicidal activity. Pesticides such as insecticides (e.g., neonicotinoïds and organophosphates such as malathion), fungicides such as mancozeb, and recently developed and introduced biopesticides such as *Bacillus thuringiensis* are not discussed in this book.

Acute toxicity tests demonstrated that glyphosate is harmless for humans, and even upon the ingestion of large doses following suicide attempts, gastroenteric discomfort was the greatest effect. For one salt form, the trimethylsulfonium or trimesium salt of glyphosate, high acute toxicity was found due to the salt counterion, but the trimesium salt of glyphosate was not included in the latest renewal procedure.

Long-term toxicity experiments in rodents in 2-year carcinogenicity testing are discussed in Chaps. 3 and 9. The use of the herbicide was further stimulated when glyphosate-resistant variants of valuable crops were synthesized. Hence, the ability to kill weeds without killing valuable crops and without acute human toxicity makes glyphosate a "once-in-a-century" herbicide (Duke, 2018; Duke & Powles, 2007, 2008).

Glyphosate and Its Metabolite Are Everywhere in All Compartments of the Environment

The use of glyphosate increased almost exponentially when Monsanto commercialized glyphosate-resistant soybean in 1996 and glyphosate-resistant corn in 1998 via genetic engineering techniques. Currently, more than 90% of soybean, cotton, and corn plants grown in the USA are glyphosate-resistant.

It is difficult to obtain an accurate literature-based view of the amount of glyphosate used currently worldwide or in different parts of the world. Peer-reviewed and reproduced data are available until 2020; most actual information is of commercial origin, which is probably correct but not peer-reviewed. The "verified market research," which is a private company, has published recent data on the glyphosate market (https://www.verifiedmarketresearch.com/product/glyphosate-market/#:~:text=Glyphosate%20Market%20Size%20And%20Forecast,5.3%25%20from%20 2023%20to%202030), showing an estimated market share increase of 5.3%, from 9.6 billion USD in 2023 to 14.7 billion USD in 2030. Similarly, Phillips and McDougall reported on glyphosate use, but these data were obtained through surveys among buyers in 43 nations. Phillips and McDougall is a private company in the UK, specializing in agribusiness which delivers predominantly economic information.

Glyphosate is found in almost all compartments of the environment, including soil, groundwater, air, and rain (Bai & Ogbourne, 2016; Chang et al., 2011; Gill et al., 2017; Hayes & Hansen, 2017; Helander et al., 2012; Zaller et al., 2022). It is found in food and drinks (Bohn & Millstone, 2019; Bou-Mitri et al., 2022; Jansons et al., 2018; Nagatomi et al., 2013; Nguyen et al., 2022; Pflaum et al., 2016; Soares et al., 2021; Vicini et al., 2021; Zoller et al., 2018), as well as in wine and beer (Garcia-Jaramillo et al., 2020; Rubio et al., 2016; Wimmer et al., 2020). Exposure in humans is evidenced by the presence of glyphosate and/or its metabolite in the urine of adults, as determined in biomonitoring studies (Andersen et al., 2022; Buekers et al., 2022a; Campbell et al., 2022; Connolly et al., 2020; De Troeyer et al., 2022; Faniband et al., 2021; Ferreira et al., 2021; Gillezeau et al., 2019; Lemke et al., 2021; Niemann et al., 2015; Nova et al., 2020; Ruiz et al., 2021; Zoller et al., 2020). Glyphosate is present in 99% of French people (Grau et al., 2022), and there are few reasons to assume that this is not the case elsewhere. Glyphosate is found in the urine of children (Buekers et al., 2022b; Ferreira et al., 2021; Knudsen et al., 2017; Lemke et al., 2021) and children living in the vicinity of glyphosate sprayed on agricultural land (Boedeker et al., 2020; Cech et al., 2023; De Troeyer et al., 2022; Ospina et al., 2022). Glyphosate is found in human hair (Alvarez et al., 2022) and in the urine of cats and dogs in New York (Karthikraj & Kannan, 2019).

One important word of caution. Chemical analysis of glyphosate is presented in most of these papers as an acquired skill and as almost routine. Researchers experienced in analytical chemistry confirm that the estimation of glyphosate in any matrix is not simple. All the data, particularly the data on concentrations near the detection limits, should be interpreted carefully. The problems encountered by the analytical assay of glyphosate have been described (Johnson et al., 2021).

Hypothetically, glyphosate and AMPA concentrations should coincide. Schwientek et al. (2024) investigated the concentrations of glyphosate and AMPA in European rivers and compared them with data from the USA. Like many others, they found glyphosate and AMPA to be present in measurable concentrations in rivers but concluded that the variability over the year was incompatible with the seasonable use of glyphosate in agriculture, leading to the suggestion of an unknown source of glyphosate. Furthermore, the variability in glyphosate concentration across seasons was not mirrored by a similar variability in AMPA. The possibility that AMPA was formed from glyphosate without enzymatic activity was proposed, and on the basis of an older paper (Klinger et al., 1998), it was suggested that atmospheric ozone might cause the reaction. However, the atmospheric concentration of ozone is usually too low for any significant reaction, as demonstrated by Klinger et al. If the data of Schwientek et al. are confirmed, then a thorough search for additional discharge of glyphosate is indicated.

Glyphosate has (almost!) gained the status of universally, omnipresent manmade molecules comparable to dioxines, chemicals such as polyfluor alkyl (PFAS group) chemicals, phthalates, bisphenols, and others. Glyphosate is not a forever chemical because of the differences in the half-lives of these molecules. Some of them remain very long in soil—dioxins: 9–15 years, DDT: 2–15 years; molecules of the PFAS group: 2–3 years up to 1000 years—whereas others, such as phthalates, degrade much more quickly, within 20 days to 2 years. The half-life of glyphosate is estimated to be between 2 and 200 days, which makes it biodegradable to some degree. Glyphosate is bound to particles in soil, and this binding is the cause of delayed metabolism. All of these molecules have—at least—a dubious and comparable reputation, not being acutely toxic but as possible low-grade toxic effectors at different physiological targets and of human function.

Taken together, the chemical properties of the molecule glyphosate and the biochemistry of its interaction with the target are fairly well known. Its excellent herbicidal properties made glyphosate the most used herbicide worldwide. Consequently, it is found in every compartment of the environment. The consequences hereof are discussed in further chapters in this book.

References

Alvarez, J. C., Etting, I., & Larabi, I. A. (2022). Glyphosate and aminomethylphosphonic acid in human hair quantified by an LC-MS/MS method. *Biomedical Chromatography, 36*(8), 11. https://doi.org/10.1002/bmc.5391

Andersen, H. R., Rambaud, L., Riou, M., Buekers, J., Remy, S., Berman, T., & Govarts, E. (2022). Exposure levels of pyrethroids, chlorpyrifos and glyphosate in EU-an overview of human biomonitoring studies published since 2000. *Toxics, 10*(12), 37. https://doi.org/10.3390/toxics10120789

Bai, S. H., & Ogbourne, S. M. (2016). Glyphosate: Environmental contamination, toxicity and potential risks to human health via food contamination. *Environmental Science and Pollution Research, 23*(19), 18988–19001. https://doi.org/10.1007/s11356-016-7425-3

Boedeker, W., Watts, M., Clausing, P., & Marquez, E. (2020). The global distribution of acute unintentional pesticide poisoning: Estimations based on a systematic review. *BMC Public Health, 20*(1), 19. https://doi.org/10.1186/s12889-020-09939-0

Bohn, T., & Millstone, E. (2019). The introduction of thousands of tonnes of glyphosate in the food chain-an evaluation of glyphosate tolerant soybeans. *Foods, 8*(12), 14. https://doi.org/10.3390/foods8120669

Bou-Mitri, C., Mekanna, A. N., Dagher, S., Moukarzel, S., & Farhat, A. (2022). Occurrence and exposure to glyphosate present in bread and flour products in Lebanon. *Food Control, 136*, 7. https://doi.org/10.1016/j.foodcont.2022.108894

Buekers, J., Remy, S., Bessems, J., Govarts, E., Rambaud, L., Riou, M., et al. (2022a). Glyphosate and AMPA in human urine of HBM4EU-aligned studies: Part B adults. *Toxics, 10*(10), 11. https://doi.org/10.3390/toxics10100552

Buekers, J., Remy, S., Bessems, J., Govarts, E., Rambaud, L., Riou, M., et al. (2022b). Glyphosate and AMPA in human urine of HBM4EU aligned studies: Part A children. *Toxics, 10*(8), 14. https://doi.org/10.3390/toxics10080470

Campbell, G., Mannetje, A., Keer, S., Eagleshama, G., Wang, X., Lin, C. Y., et al. (2022). Characterization of glyphosate and AMPA concentrations in the urine of Australian and New Zealand populations. *Science of the Total Environment, 847*, 5. https://doi.org/10.1016/j.scitotenv.2022.157585

Cech, R., Zaller, J. G., Lyssimachou, A., Clausing, P., Hertoge, K., & Linhart, C. (2023). Pesticide drift mitigation measures appear to reduce contamination of non-agricultural areas, but hazards to humans and the environment remain br. *Science of the Total Environment, 854*, 11. https://doi.org/10.1016/j.scitotenv.2022.158814

Chang, F. C., Simcik, M. F., & Capel, P. D. (2011). Occurrence and fate of the herbicide glyphosate and its degradate aminomethylphosphonic acid in the atmosphere. *Environmental Toxicology and Chemistry, 30*(3), 548–555. https://doi.org/10.1002/etc.431

Connolly, A., Koslitz, S., Bury, D., Bruning, T., Conrad, A., Kolossa-Gehring, M., et al. (2020). Sensitive and selective quantification of glyphosate and aminomethylphosphonic acid (AMPA) in urine of the general population by gas chromatography-tandem mass spectrometry. *Journal of Chromatography B-Analytical Technologies in the Biomedical and Life Sciences, 1158*, 5. https://doi.org/10.1016/j.jchromb.2020.122348

De Troeyer, K., Casas, L., Bijnens, E. M., Bruckers, L., Covaci, A., De Henauw, S., et al. (2022). Higher proportion of agricultural land use around the residence is associated with higher urinary concentrations of AMPA, a glyphosate metabolite. *International Journal of Hygiene and Environmental Health, 246*, 8. https://doi.org/10.1016/j.ijheh.2022.114039

Duke, S. O. (2018). The history and current status of glyphosate. *Pest Management Science, 74*(5), 1027–1034. https://doi.org/10.1002/ps.4652

Duke, S. O., & Powles, S. B. (2007). AGRO 1-Glyphosate: A once-in-a-century herbicide. *Abstracts of Papers of the American Chemical Society, 233*, 302–302. Retrieved from <Go to ISI>://WOS:000207722800277.

Duke, S. O., & Powles, S. B. (2008). Glyphosate: A once-in-a-century herbicide. *Pest Management Science, 64*(4), 319–325. https://doi.org/10.1002/ps.1518

Faniband, M. H., Noren, E., Littorin, M., & Lindh, C. H. (2021). Human experimental exposure to glyphosate and biomonitoring of young Swedish adults. *International Journal of Hygiene and Environmental Health, 231*, 6. https://doi.org/10.1016/j.ijheh.2020.113657

Ferreira, C., Duarte, S. C., Costa, E., Pereira, A., Silva, L. J. G., Almeida, A., et al. (2021). Urine biomonitoring of glyphosate in children: Exposure and risk assessment. *Environmental Research, 198*, 7. https://doi.org/10.1016/j.envres.2021.111294

Garcia-Jaramillo, M., Spokas, K. A., Levin, A. D., Osborne, J., Acosta-Martinez, V., & Trippe, K. M. (2020). Soil phosphorus content determines grapevine glyphosate uptake and accumulation in wine. *Hortscience, 55*(9), S317–S317. Retrieved from <Go to ISI>://WOS:000623253400646.

Gill, J. P. K., Sethi, N., & Mohan, A. (2017). Analysis of the glyphosate herbicide in water, soil and food using derivatising agents. *Environmental Chemistry Letters, 15*(1), 85–100. https://doi.org/10.1007/s10311-016-0585-z

Gill, J. P. K., Sethi, N., Mohan, A., Datta, S., & Girdhar, M. (2018). Glyphosate toxicity for animals. *Environmental Chemistry Letters, 16*(2), 401–426. https://doi.org/10.1007/s10311-017-0689-0

Gillezeau, C., van Gerwen, M., Shaffer, R. M., Rana, I., Zhang, L. P., Sheppard, L., & Taioli, E. (2019). The evidence of human exposure to glyphosate: A review. *Environmental Health, 18*, 14. https://doi.org/10.1186/s12940-018-0435-5

Grau, D., Grau, N., Gascuel, Q., Paroissin, C., Stratonovitch, C., Lairon, D., et al. (2022). Quantifiable urine glyphosate levels detected in 99% of the French population, with higher values in men, in younger people, and in farmers. *Environmental Science and Pollution Research, 29*(22), 32882–32893. https://doi.org/10.1007/s11356-021-18110-0

Hayes, T. B., & Hansen, M. (2017). From silent spring to silent night: Agrochemicals and the anthropocene. *Elementa-Science of the Anthropocene, 5*, 24. https://doi.org/10.1525/elementa.246

Helander, M., Saloniemi, I., & Saikkonen, K. (2012). Glyphosate in northern ecosystems. *Trends in Plant Science, 17*(10), 569–574. https://doi.org/10.1016/j.tplants.2012.05.008

Jansons, M., Pugajeva, I., & Bartkevics, V. (2018). Occurrence of glyphosate in beer from the Latvian market. *Food Additives and Contaminants Part a-Chemistry Analysis Control Exposure & Risk Assessment, 35*(9), 1767–1775. https://doi.org/10.1080/19440049.2018.1469051

Johnson, C. L., Jazan, E., Kong, S. W., & Pennell, K. D. (2021). A two-step gas chromatography-tandem mass spectrometry method for measurement of multiple environmental pollutants in human plasma. *Environmental Science and Pollution Research, 28*(3), 3266–3279. https://doi.org/10.1007/s11356-020-10702-6

Karthikraj, R., & Kannan, K. (2019). Widespread occurrence of glyphosate in urine from pet dogs and cats in New York State, USA. *Science of the Total Environment, 659*, 790–795. https://doi.org/10.1016/j.scitotenv.2018.12.454

Klinger, J., Lang, M., Sacher, F., Brauch, H. J., Maier, D., & Worch, E. (1998). Formation of glyphosate and AMPA during ozonation of waters containing ethylenediaminetetra(methylenephosphonic acid). *Ozone-Science & Engineering, 20*(2), 99–110. https://doi.org/10.1080/01919519808547279

Knudsen, L. E., Hansen, P. W., Mizrak, S., Hansen, H. K., Morck, T. A., Nielsen, F., et al. (2017). Biomonitoring of Danish school children and mothers including biomarkers of PBDE and glyphosate. *Reviews on Environmental Health, 32*(3), 279–290. https://doi.org/10.1515/reveh-2016-0067

Leino, L., Tall, T., Helander, M., Saloniemi, I., Saikkonen, K., Ruuskanen, S., & Puigbò, P. (2021). Classification of the glyphosate target enzyme (5-enolpyruvylshikimate-3-phosphate synthase) for assessing sensitivity of organisms to the herbicide. *Journal of Hazardous Materials, 408*, 8. https://doi.org/10.1016/j.jhazmat.2020.124556

Lemke, N., Murawski, A., Schmied-Tobies, M. I. H., Rucic, E., Hoppe, H. W., Conrad, A., & Marike, G. (2021). Glyphosate and aminomethylphosphonic acid (AMPA) in urine of children and adolescents in Germany – Human biomonitoring results of the German Environmental Survey 2014–2017 (GerES V). *Environment International, 156*, 9. https://doi.org/10.1016/j.envint.2021.106769

Maeda, H., & Dudareva, N. (2012). The shikimate pathway and aromatic amino acid biosynthesis in plants. In S. S. Merchant (Ed.), *Annual review of plant biology* (Vol. 63, pp. 73–105). Annual Reviews.

Mesnage, R., Benbrook, C., & Antoniou, M. N. (2019). Insight into the confusion over surfactant co-formulants in glyphosate-based herbicides. *Food and Chemical Toxicology, 128*, 137–145. https://doi.org/10.1016/j.fct.2019.03.053

Nagatomi, Y., Yoshioka, T., Yanagisawa, M., Uyama, A., & Mochizuki, N. (2013). Simultaneous LC-MS/MS analysis of glyphosate, glufosinate, and their metabolic products in beer, barley tea, and their ingredients. *Bioscience Biotechnology and Biochemistry, 77*(11), 2218–2221. https://doi.org/10.1271/bbb.130433

Nguyen, M. H., Nguyen, T. D., Vu, M. T., Duong, H. A., & Pham, H. V. (2022). Determination of glyphosate, glufosinate, and their major metabolites in tea infusions by dual-channel capillary electrophoresis following solid-phase extraction. *Journal of Analytical Methods in Chemistry, 2022*, 9. https://doi.org/10.1155/2022/5687025

Niemann, L., Sieke, C., Pfeil, R., & Solecki, R. (2015). A critical review of glyphosate findings in human urine samples and comparison with the exposure of operators and consumers. *Journal Fur Verbraucherschutz Und Lebensmittelsicherheit-Journal of Consumer Protection and Food Safety, 10*(1), 3–12. https://doi.org/10.1007/s00003-014-0927-3

Nova, P., Calheiros, C. S. C., & Silva, M. (2020). Glyphosate in Portuguese adults – A pilot study. *Environmental Toxicology and Pharmacology, 80*, 6. https://doi.org/10.1016/j.etap.2020.103462

Ospina, M., Schutze, A., Morales-Agudelo, P., Vidal, M., Wong, L. Y., & Calafat, A. M. (2022). Exposure to glyphosate in the United States: Data from the 2013–2014 National Health and Nutrition Examination Survey. *Environment International, 170*, 8. https://doi.org/10.1016/j.envint.2022.107620

Pflaum, T., Hausler, T., Baumung, C., Ackermann, S., Kuballa, T., Rehm, J., & Lachenmeier, D. W. (2016). Carcinogenic compounds in alcoholic beverages: An update. *Archives of Toxicology, 90*(10), 2349–2367. https://doi.org/10.1007/s00204-016-1770-3

Rubio, F., Hutchinson, Z., Glaze, T., & Lance, J. (2016). Survey of glyphosate in domestic and imported beer and wine. *Abstracts of Papers of the American Chemical Society, 252*, 1. Retrieved from <Go to ISI>://WOS:000431460200599.

Ruiz, P., Dualde, P., Coscolla, C., Fernandez, S. F., Carbonell, E., & Yusa, V. (2021). Biomonitoring of glyphosate and AMPA in the urine of Spanish lactating mothers. *Science of the Total Environment, 801*, 7. https://doi.org/10.1016/j.scitotenv.2021.149688

Schwientek, M., Rügner, H., Haderlein, S. B., Schulz, W., Wimmer, B., Engelbart, L., et al. (2024). Glyphosate contamination in European rivers not from herbicide application? *Water Research, 263*, 10. https://doi.org/10.1016/j.watres.2024.122140

Soares, D., Silva, L., Duarte, S., Pena, A., & Pereira, A. (2021). Glyphosate use, toxicity and occurrence in food. *Foods, 10*(11), 22. https://doi.org/10.3390/foods10112785

Travlos, I., Cheimona, N., & Bilalis, D. (2017). Glyphosate efficacy of different salt formulations and adjuvant additives on various weeds. *Agronomy-Basel, 7*(3), 9. https://doi.org/10.3390/agronomy7030060

Vicini, J. L., Jensen, P. K., Young, B. M., & Swarthout, J. T. (2021). Residues of glyphosate in food and dietary exposure. *Comprehensive Reviews in Food Science and Food Safety, 20*(5), 5226–5257. https://doi.org/10.1111/1541-4337.12822

Wimmer, B., Pattky, M., Zada, L. G., Meixner, M., Haderlein, S. B., Zimmermann, H. P., & Huhn, C. (2020). Capillary electrophoresis-mass spectrometry for the direct analysis of glyphosate: Method development and application to beer beverages and environmental studies. *Analytical and Bioanalytical Chemistry, 412*(20), 4967–4983. https://doi.org/10.1007/s00216-020-02751-0

Zaller, J. G., Kruse-Plass, M., Schlechtriemen, U., Gruber, E., Peer, M., Nadeem, I., et al. (2022). Pesticides in ambient air, influenced by surrounding land use and weather, pose a potential threat to biodiversity and humans. *Science of the Total Environment, 838*, 12. https://doi.org/10.1016/j.scitotenv.2022.156012

Zoller, O., Rhyn, P., Rupp, H., Zarn, J. A., & Geiser, C. (2018). Glyphosate residues in Swiss market foods: Monitoring and risk evaluation. *Food Additives & Contaminants Part B-Surveillance, 11*(2), 83–91. https://doi.org/10.1080/19393210.2017.1419509

Zoller, O., Rhyn, P., Zarn, J. A., & Dudler, V. (2020). Urine glyphosate level as a quantitative biomarker of oral exposure. *International Journal of Hygiene and Environmental Health, 228*, 6. https://doi.org/10.1016/j.ijheh.2020.113526

Chapter 2
Introductory Notes on Diseases

This chapter presents some basic tenets on toxicology, the differences between risk assessment and risk management, and epidemiology. Because of its importance, the precautionary principle as a management tool is discussed in more detail. As mentioned in the general introduction, completeness is not intended; the interested reader is referred to the appropriate extensive literature and textbooks.

Toxicity Testing, Animal Experiments

The introduction on the market of new chemicals that potentially come in contact with consumers, such as food additives, pharmaceuticals, and pesticides, is due to the strict control of toxicity. To date, the legally required tests are all animal tests, mostly in rodents (mice and rats) and rarely in primates. For particular tests, e.g., skin testing, rabbits are frequently used.

The three main groups of tests are acute toxicity, mid- to long-term toxicity after repeated dosing, and long-term toxicity after 24 months of dosing.

Acute toxicity, known as "the six-pack," consists of three systemic tests: oral toxicity, during which the animal is observed for 24 hours after one single exposure; acute inhalation toxicity during which the animal is observed for 4 hours; and acute dermal toxicity during which the testing time window is 24 hours. In addition, three organ-specific tests are needed: primary eye irritation, primary skin irritation, and dermal sensitization. Mid-/long-term toxicity testing includes repeated-dose toxicity testing for 28 days and a sub-chronic 90-day test with daily oral administration. The long-term toxicity test consists of 24 months of carcinogenicity testing with daily administration of the compound via food and water. See Parasuraman for a full description (https://journals.sagepub.com/doi/abs/10.4103/0976-500X.81895).

In addition to these classical tests, some additional experimental results are required.

N. Fraeyman, *Glyphosate 2023–2033*,
https://doi.org/10.1007/978-3-031-97540-0_2

Transgenerational testing includes the follow-up of the toxic phenotype in the first (F1) and second (F2) generations of dams treated with the compound.

If a compound is suspected of carcinogenicity, the 2-year animal tests are supplemented with a number of in vitro tests such as the AMES test, sister chromatid exchange test, and comet assay. The AMES test is named after Bruce Ames, who introduced the test in the 70ies. This biological assay detects the mutagenic capacity of chemicals using mutations in the DNA of specific strains of Salmonella bacteria. In the sister chromatid exchange test, the exchange of genetic material between sister chromatids of a chromosome following DNA damage is taken as a sign of the mutagenic properties of the chemical. The comet test is a single cell electrophoretic assay in which the extent of the migration through the gel of fragments of DNA which forms the shape of a comet, is representative of the DNA damage.

In general, compounds can be carcinogenic, i.e., exposure of animals and people could result in cancer. They can be mutagenic, i.e., exposure can cause mutations of the DNA of cells, although this damage does not necessarily lead to overt cancer. Carcinogens do not necessarily have to interact with DNA but might have other cellular targets (see Chap. 3 on cancer mechanisms). They can be teratogens or embryo toxins; i.e., they can cause birth defects, developmental delays, or death in offspring. They can be reproductive toxins, i.e., they can cause sterility, reduced fertility, and semen toxicity.

The protocols for performing toxicity tests are described meticulously, and the OECD guidelines are considered the gold standard. This is a collection of more than 150 internationally agreed-upon testing methods followed by industry, governments, and independent laboratories (https://www.oecd-ilibrary.org/environment/oecd-guidelines-for-the-testing-of-chemicals_72d77764-en).

To give one example (not in full detail). Two-year carcinogenicity testing requires groups of rodents to be treated for 2 years with the compound in their food and water. As a rule, 3–5 doses in addition to a blank (nontreated) group are used, and 50 female and 50 male animals are needed per dose. For example, for three doses, 400 mice are needed, and these animals need to be "young" in animal terms (a few weeks) since mice or rats usually do not live much longer than 2 years. Rodents are followed on a daily basis for basic parameters such as survival, body weight, food, and water consumption and for clinical symptoms such as respiratory symptoms and behavior. At the end of the experiments, after 24 months, the animals are sacrificed, and the organs are analyzed for deviations from the organs of the control animals (weight, color, abnormal growth, etc.). Histopathological analysis is used to detect the eventual presence of neoplasms or precancerous lesions, distinguish benign or malignant lesions, etc. Clinical chemistry analysis of the blood yields information, e.g., liver and kidney function, glucose, lipids, etc., which is very much comparable to that of human clinical chemistry routine analysis. The full package of results, including the dose–response relationship, provides a complete overview of the possible toxicity of the compound. There is no need to stress the high costs and complicated logistics of this type of experiment.

In earlier days, these experiments were performed by the company that produces the chemical, often not according to the rules of good practice, and on top of that, the results are frequently manipulated, twisted, or not reported (McHenry, 2018). Currently, these experiments are mostly performed by dedicated laboratories with considerable experience. The results, however, remain the property of the paying organization, which can, theoretically, use them as wanted.

Animal testing is highly controversial, and numerous attempts to replace animals with in vitro toxicity tests have been published (Benigni et al., 2013, 2015; Benigni & Bossa, 2011; Knight et al., 2006a, b, c; Maertens et al., 2024b; Zhang et al., 2010). The future of toxicity testing depends on methodological developmental steps such as high-throughput in vitro screening, computational toxicology, functional genomics, transcriptomics, and genomics. The consequences for human health have been discussed (Krewski et al., 2009). The Belgian Superior Health Council has recently published advice on state-of-the-art on genomics and its use in clinical practice and the integration of genomics in daily practices, including toxicity testing https://www.health.belgium.be/sites/default/files/uploads/fields/fpshealth_theme_file/20210203_hgr-9477_omics_vweb.pdf. Artificial intelligence and data-driven methods are likely to complement or even replace older animal testing methods (Beilmann et al., 2019; Hartung, 2019a; Maertens et al., 2022, 2024a; Russo et al., 2019). Furthermore, the predictability of animal tests is not waterproof, and with sophisticated software, similar predictability was obtained with an in vitro approach (Hartung, 2019b).

One major problem with a 2-year study is that the animals are very old at the end of the study, and a portion of the 50 original mice die before the end of the study. Similar to other animals, including men, mice and rats develop cancer as they age. If a compound is carcinogenic, the number and type of compound-induced tumors must be compared to those of naturally growing cancers, which can jeopardize the statistical outcome.

Risk Assessment and Risk Management, Quantification of the Exposure

Risk and hazard are frequently used together, but they are not interchangeable.

*Risk is a quantitative/***statistical variable** *expressed as the possibility of personal suffering, e.g., from a disease or from property loss per any feasible unit, (e.g., unit of time or number of potential victims). The Cambridge definition of risk is "the possibility of something bad happening". "Hazard is an* **intrinsic property** *of the situation, chemical, pathogen, etc. The link between risk and hazard is exposure.*

Risk occurs on a daily basis but is, in many cases, neglected. Statistically, 1–10 fatal accidents per 200 million km on the road occur, but when we drive our car, we hardly care about this risk. This is the case in many daily activities.

Hazards are frequently unknown, in particular for many chemicals. Even if the hazard is known or suspected, it is often neglected, difficult to evaluate or

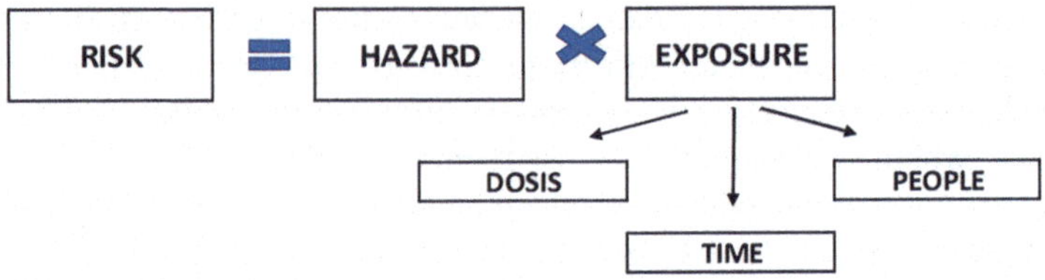

Fig. 2.1 Relationships among risk, hazard, and exposure

considered not important. This is frequently the case when a negative effect of exposure is only observable after a latency period. The carcinogenic effect of asbestos appears after a latency period of 3–5 decades, which makes it difficult to attribute the negative effect to the exposure.

Exposure depends on some variables, with the most important being the duration, the intensity expressed as the number of doses per unit of time (or any other unit), and the bodily constitution of the potential victim (e.g., degree of vulnerability). Complicating factors depending on the specific item might interfere, e.g., the interaction between the dose and duration of exposure. This relation is conveniently depicted in Fig. 2.1.

The hazard might be (very) high, but without exposure, there is no risk. Conversely, a moderate to low quantitative hazard might lead to devastating consequences when the dose of the exposure is high or long.

In practice, the risk-hazard-exposure triplet can be difficult to evaluate. For many substances, the hazard is not known. Wang et al. (2020) reported on 22 chemical inventories in 19 countries and found over 350,000 chemicals of which 50,000 are publicly unknown because of confidentiality and 70,000 are ambiguously described. Nearly 240,000 chemicals have a CAS number, but only a minor fraction has been investigated for hazard, let alone for risk. In Europe, the registration of chemicals is according to rules within REACH (Registration, Evaluation, Authorisation and Restriction of Chemicals), provided that the volume produced by the company is more than 1 ton per year. Currently, approximately 22,000 chemicals are registered, and approximately 200 are considered very high-concern substances. In case of doubt, the precautionary principle could (or should) apply.

Toxicity testing leads to an approximation of the risk for human health upon exposure. This exposure is quantitatively expressed as the allowable daily intake (ADI—chemicals such as food additives that are deliberately added) or the tolerable daily intake (TDI—extracorporeal chemicals with possible health-disturbing properties). Both are lifetime threshold values, expressed in mg/kg body weight per day, considered safe for humans. The values of ADI or TDI are derived from the results of the toxicity testing experiments in which multiple doses of the compound were used to construct a dose–response curve for measured effects. Among these doses, the lowest dose in the series that has an effect is called the lowest observable effect level (LOEL), and the next lower dose that has no effect is called the no observable effect level (NOEL). The effects or response are very diverse and depend on the chemical and its biological target. Examples include alterations in enzyme

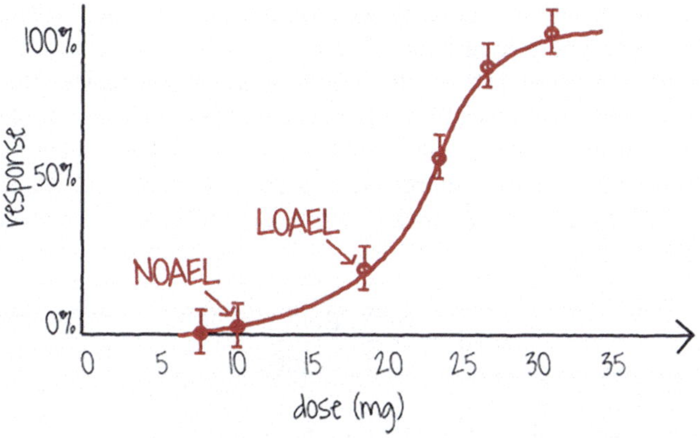

Fig. 2.2 Theoretical example of a dose–response curve in toxicity testing experiments

activity in the blood, liver damage, kidney damage, reduced weight, fetal malformation, seizures, as well as motor skills and behavior. Some of these effects are permanent, whereas others are reversible. In these cases, the NOEL and LOEL are replaced by the no observable adverse effect level (NOAEL) and the lowest observable adverse effect level (LOAEL), respectively. This is shown in Fig. 2.2.

The LO(A)EL is used to calculate the ADI (or TDI) by including a safety factor or uncertainty factor, allowing extrapolation from animals to humans. This safety factor can be any number but is frequently set to 100 (Ku, 2001).

$$\mathrm{ADI} = \frac{\mathrm{NOAEL}}{\mathrm{SAFETY\ FACTOR}}$$

The current ADI values for glyphosate are: 0.3 mg/kg/day for the World Health Organisation (WHO), 0.5 mg/kg/day for EFSA, and 1.75 mg/kg/day for the American Environmental Protection Agency (EPA).

The whole method has been criticized many times because of the arbitrariness of the safety factor; this is also called the "uncertainty factor" illustrating its scientific insecurity (Landis, 2012). The commonly used factor of 100 is derived form a factor of 10 for interspecies variability multiplied by 10 for intraspecies variability. If the uncertainty is more pronounced due to e.g. missing data, a factor of 1000 could be applied. The alternative to using a safety factor is curve fitting, whereby the experimental dose–response curves are extended toward the lower concentration range. Curve fitting would allow a more robust statistical analysis of NOAEL or LOAEL and an accurate estimation of ADI or TDI. This possibility is not further elaborated in this book.

The problem of limiting exposure is complicated for carcinogens since it is impossible to establish a safe exposure limit; even small doses might lead to cancerous diseases (Nohmi, 2018). Instead of a no-threshold model, other approaches, such as a risk assessment model (e.g., possible number of cancers per one million people) are used (Högberg & Järnberg, 2023).

Determining the ADI or TDI is important to manufacturers since this has implications for their use in daily life. The link between ADI and a safe dose (at least for rodents) is the consumption of food/drinks per day. For the human situation, the question is how much of the compound is present in food and drinks, which is expressed as the MRL or maximum residue level, and how much of the food and drinks we consume.

$$\text{MRL} = \frac{\text{ADI} \times \text{Body weight}}{\text{Food intake}}$$

MRL is the amount of the substance tolerable in the consumable items and is expressed in mg/kg; the body weight is frequently taken as 60–70 kg for an adult person, and food intake is expressed as kg/day.

The MRL is the key variable since it is the amount of pesticide present in the cultures at harvest after the application of the compound. As a rule, MRL values among all crops that are treated with the compound are very variable, and their estimation is required for the application. Establishing the ADI value for a pesticide is deterministic for the applicability of the pesticide, i.e., the amount of pesticide that can be sprayed on the crops.

For practically all food, fruit, vegetables, and meat, the MRL is a fixed value. These values can be found at: https://ec.europa.eu/food/plant/pesticides/eu-pesticides-database/start/screen/mrls/details?lg_code=EN&pest_res_id_list=120,223&product_id_list=.

As a rule, and in the EU, the MRL of glyphosate is between 0.1 and 10 mg/kg in fruits, between 0.1 and 0.5 mg/kg in vegetables, 0.1 mg/kg in wine, and 0.01 mg/kg in meat, including poultry. This is discussed in detail by the EFSA in 2019 (https://efsa.onlinelibrary.wileu.com/doi/epdf/10.2903/j.efsa.2019.5862).

Taken together, a lot of efforts are made to describe the average exposure of people to glyphosate leading to values for the ADI, and nowadays, sophisticated software allows rigorous statistical analysis. The UK Health and Security Agency recently published its "Guidance to Risk Characterisation Methods" (22/8/2022), which gives a detailed overview of the possible methods and pitfalls of quantification of risk assessment methods (https://www.gov.uk/government/publications/cancer-risk-characterisation-methods/risk-characterisation-methods).

Of note, ADI, TDI, and MRL are some of the many abbreviations used in this context (see Chapter 11, Annex 1).

Epidemiology, General Information

Information on environmental human toxicology is obtained mainly from epidemiological studies, and the intrinsic difficulties of this type of study are well known (https://www.ippnw.org/wp-content/uploads/2023/03/MGSV1N2Wing.pdf).

Epidemiology is a valid scientific discipline with an extended and sophisticated number of armaments at its disposal for corrections of bias, confounders, etc., including advanced statistical methodology (Lash et al., 2021). The main problem in population-wide epidemiological studies is that individual diseases are recorded, but individual exposure to environmental pollutants is not. Hence, the cause–consequence issue or any conclusion on a direct one-on-one relationship is impossible making the link between epidemiological data and human risk difficult (de Almeida et al., 2009), and this missing information is eagerly used by proponents of the further use of chemicals.

The situation is complex because of two additional variables.

First, the occurrence of diseases is not only due to the negative influence of the environment but also dependent on the genetic makeup of every individual (Münzel et al., 2021). The argument that the environment is "not to blame" for the prevalence of diseases but is "only a partner" is frequently used to shift the guild from a polluting chemical to the genetics of the victim. This argument is only partly valid and is not deterministic. Many diseases are caused by changes in the genetics of the victim (10–30%—https://www.britannica.com/science/human-genetic-disease), including mutations and deletions, and many of these alterations in the basic genetic profile are due to environmental aggressors (Virolainen et al., 2023). Hence, the argument that genetics is more important than the environment is faulty.

Second, becoming a patient is a statistic. The risk of lung cancer is approximately 15 (for males) and 20 (for females) times greater in smokers than in non-smokers, indicating that non-smokers might also get lung cancer while smokers might remain healthy. This observation is used frequently to contest the outcome of epidemiological data, not only on smoking. This kind of criticism is used by people who are very familiar with statistics and know how to (mis)use and interpret data to support their views.

The epidemiological data on cancer incidence and deaths, although important, are only partially useful for studying the link between the environment and health for at least three reasons: (i) There is a highly variable latency period between exposure and overt expression of cancer, obscuring the interpretation of the time sequence of exposure and disease (https://www.cdc.gov/wtc/pdfs/policies/WTCHP-Minimum-Cancer-Latency-PP-01062015-508.pdf). (ii) The evolution of the number of cancer deaths as a criterion for exposure to environmental pollutants is not useful given the significant amelioration of the control over the illness during the last decades (luckily for many types of cancers, unfortunately not for all). (iii) People are never exposed to one polluting chemical or extracorporeal stress; this item is discussed in Chap. 7. The outcome is always the result of multiple exposures, making the identification of one single culprit near to impossible.

Conclusions on health problems due to exposure to polluting chemicals frequently clash on this cause–consequence issue. One possibility to overcome this relies on the application of the Bradford–Hill criteria (Hill, 1965). In principle, the single criterion on cause–consequence is replaced by nine criteria covering other variables. These nine criteria include (1) evidence for the association between exposure and disease, (2) consistency of the link between the location of the environmental pollution and the affected population, (3) specificity of the disease with the type of environmental challenge, (4) sequence or temporality, meaning that the disease must come after the exposure, (5) the existence of a dose–response or biological gradient or at least an indication, (6) biological evidence or plausibility of the biomedical consequence of the exposure, (7) coherence that the biomedical consequence must make sense, (8) cause–consequence should be proven experimentally, eventually in animals or in the laboratory, and (9) similar agents should cause similar diseases.

Although nearly 90 years old, these criteria are still used (Cox, 2018; Gross & Fedak, 2015; Shimonovich et al., 2021; Weed, 2018). In many cases, not all criteria are fulfilled because of missing data since experiments are not possible for evident ethical reasons. In practice, some of the nine criteria are prioritized, including temporality (consequence or disease must follow the exposure to the pollutant), consistency (equal exposure leads to an equal outcome in, e.g., different locations), chance (when the exposure stops, the disease stops), and strength of the relationship (biological evidence for the effect and dose–response relationship). Here, again, missing data are frequently cited as important, flagrantly neglecting convincing other epidemiological data. Notably, Bradford-Hill never used the term "criteria" for the nine conditions, but instead called them "viewpoints". Recently, a discussion on this point was published (https://pubmed.ncbi.nlm.nih.gov/30433840).

Who Is at Risk? Who Is Exposed to Glyphosate?

Since glyphosate is found in every compartment of the environment, essentially everybody is unwillingly exposed, in most cases to minute amounts of the chemical. Some groups are more at risk because of potentially higher exposure than others. These include the following:

- Professionals spraying glyphosate on the fields: Although this is an evident assumption, this is rarely the case. These professionals are, as a rule, well-trained, are licensed to do the spraying, have access to all protective measures (clothing, gloves, masks, etc.), and apply the compound in a well-isolated cabin of the motorized vehicle. The exposure to glyphosate (or any other pesticide) during the spraying is probably low.
- Farmers during work on the field: Professional spraying is applied to large(r) areas, but for smaller pieces of land, farmers could have the tendency to apply the pesticide themselves. In this case, the possibility of bodily contamination is much higher. The follow-up of using protective clothing, although advised by the manufacturer, might be less strict. Hence, exposure is expected to be higher compared to professional sprayers.
- Amateur gardeners or people doing garden maintenance are more exposed. This use of glyphosate is forbidden in many EU countries. Only professional sprayers can buy formulations of glyphosate, which would eliminate this non-legal use.

- One of the important problems encountered is the unwilling spread of the herbicide during application, called "spray drift management" which causes the unwilling exposure of bystanders, as convincingly demonstrated recently (De Troeyer et al., 2022). Notably, EFSA considers the toxicity of glyphosate for bystanders unimportant since no acute toxicity is linked to the molecule, and there are, according to EFSA, negligible long-term effects. In contrast, the Environmental Protection Agency (EPA) in the USA has published stringent guidelines for professional sprayers, including not to spray during a temperature inversion or when the wind speed is over 10 mph (https://www.epa.gov/sites/default/files/2019-04/documents/glyphosate-response-comments-usage-benefits-final.pdf). The use of artificial barriers against the unwilling spread of pesticides has been promoted (Ucar & Hall, 2001) but has not yet been widely applied.

Taken together, many people are exposed to glyphosate, either during application on the fields or unwillingly due to drift during spraying. Particularly important is the exposure of the most vulnerable, such as the unborn fetus, due to the presence of glyphosate in the mother and the young growing child.

The Precautionary Principle: Dancing on a Tightrope

The conclusion from the previous chapters is that making decisions on environmental health issues is not straightforward. The following paragraphs focus on making critical decisions without having all the information you want or need. Incomplete information is the logical and irrefutable consequence of scientific research since research evolves, and new findings could confirm and complete but also contradict earlier findings. Information might be timely correct or incorrect which may lead to both false negative (too cautious) or false positive (too liberal) attitudes as far as ensuing consequences are concerned. This is where the precautionary principle comes into the picture. The focus of the next discussion is on taking decisions about human health and environmental pollution with chemicals or preparations containing chemicals without having all the information. Daily care products such as cosmetics, hygienic products, and hair coloring dyes are excluded from this discussion.

There is not one single universal definition of the precautionary principle; many descriptions are available. Principle 15 of the Rio Declaration (1992) states: "*To protect the environment, the precautionary approach shall be widely applied by States according to their capabilities. Where there are threats of serious or irreversible damage, lack of full scientific certainty shall be not used as a reason for postponing cost-effective measures to prevent environmental degradation.*" The Wingspread statement of 1998 states: "*Where an activity raises threats of harm to the environment or human health, precautionary measures should be taken even if some cause–and–effect relationships are not fully established scientifically*". The precautionary principle is freely translated in well-known one-liners: better safe than sorry or do not harm or be careful. Initially intended for environmental issues, the definition was readily used for other human health issues. The precautionary principle could be considered as a risk assessment tool, and a risk management tool.

The precautionary principle addresses primary prevention (avoiding the occurrence of an event with negative effects) and is the opposite of secondary prevention, which addresses the avoidance of the re-occurrence of the negative event. It is about the protection of human life, particularly the life of the most vulnerable, whereas this class of society is, as a rule, least involved in the decision-making process but undergoes the consequences.

The precautionary principle is about the safety in principle of all members of the society. Safety is not a dichotomous variable between safe and unsafe, but it is a continuous variable between absolute safety and the complete absence of safety. Absolute safety is beyond reach, and in the context of this principle, accepting some unavoidable remaining risk is essential. The study of risk includes risk assessment, which is a quantitative action, and risk management, which is a qualitative action that uses the data from the risk assessment and combines these data with other, possibly less quantitative information. Questions such as "What is a threat or serious damage? Or even "What is serious damage?" "What is full scientific certainty?" and "What are cost-effective measures?" have fueled thousands of papers and books and have been at the center of heated debates in lawsuits. A theoretical analysis of the principle has been described in detail previously (https://pubmed.ncbi.nlm.nih.gov/20653252/).

The application of the precautionary principle in real-life situations is difficult; there is no mathematical rule to judge whether the principle is applicable, and it certainly is not a black-or-white system. In most cases, and for chemicals, the precautionary principle is about a balance between beneficial properties to society as a whole and detrimental properties for (human) health and safety and the environment. The challenge of explaining the complexity of the principle to the European Parliament was taken up approximately 9 years ago (https://www.europarl.europa.eu/thinktank/en/document/EPRS_IDA(2015)573876) and discussed on multiple occasions (Goldstein & Carruth, 2005; Hemphill, 2020). At the EU level, the precautionary principle is one of the pillars, although misuse of the principle must be avoided: "*The precautionary principle may only be invoked if there is a potential risk and may not be used to justify arbitrary decisions*" (https://echa.europa.eu/regulations/reach/legislation for an overview of the legislation).

Introducing a chemical on the market is subject to at least four influences (probably in this order of priority): economic (the assumed profit for the manufacturer and stakeholders), societal (acceptance by the public eventually after the creation of an artificial "new need" and "fear of missing out"), efficiency or acceptance (based on a positive effect at the intended target), and the absence of unwanted and unintended off-target effects. Instead, the first question should be whether a chemical or a composite could cause health problems for humans upon deliberate or unwanted exposure. The information on human toxicology upon marketing is mostly incomplete, incorrect, or, in the worst cases, biased or kept secret and subject to unwanted influences such as lobbying, and all this is stirred up by fake news, deliberate wrong information, conspiracy theories, etc.

All this means that the results of any type of toxicology experiment are not final, and a mechanism of reviewing a piece of advice or decision whenever new information becomes available should be installed. What is imperative is to define beforehand the criteria for whether new information is relevant, and the EFSA is well aware of this (see https://www.efsa.europa.eu/en/efsajournal/pub/8312, https://www.efsa.europa.eu/en/efsajournal/pub/4737, and https://www.efsa.europa.eu/en/efsajournal/pub/5123, for the most important criteria). In principle, the review of the new information eventually leading to the re-evaluation of earlier decisions should be performed by an independent organization to prevent the EFSA from being a judge and a party. Moreover, powerful companies, organized in lobbying task forces with great technical, organizational, and intellectual freedom and possessing much more money than official, governmental, and super-governmental organizations, influence decisions. In the EU, members of the parliament must register their meetings with lobbyists when the meeting is within the perimeter of the parliament. Most meetings are evidently outside the parliament, so this rule is a sham rule to comply with administrative obligations and soothe the conscience of lobbyists and some members of the parliament. This point is also discussed in Chaps. 3 and 9.

A second problem is the wealth of information. Selection bias, i.e., using the papers/information that fit his own opinion and disregarding information that argues against the opinion, is possible. The opinions of famous and established organizations might be biased and eventually historically formed, which makes it difficult for individual scientists to overrule.

Taken together, the possibility of applying the precautionary principle in situations suffering from incomplete information is welcome. The application in practice is difficult because of multiple influences defining the balance between economic profit and human health.

This chapter is on the use of the precautionary principle in three historical examples (DDT, BPA and asbestos) and an attempt to extrapolate this principle to four new actual situations (micro- and nanoplastics, PFAS, glyphosate, and smoking). This distinction of historical/actual is somewhat artificial since some discussions, e.g., on asbestos, are still ongoing. The accepted definition of the precautionary principle and the pending legal obligations are taken as a starting point. One of the main upfront conclusions is that legal obligations are interpreted differently by different stakeholders, which is likely one of the main problems.

The use of *dichloro-diphenyl-trichloroethane, or DDT*, against malaria was described by Paul Hermann Müller, who received the Nobel Prize in 1939 for this work. Its use during the Second World War has saved many lives of soldiers who otherwise would have died of typhus, malaria, and other infections. After the war, DDT was massively used in agriculture and daily life for pest control, and very soon, the side effects were devastating and far worse than the positive effects. The killing of birds (an observation that inspired Rachel Carson for the title of her book in 1962) was the observable consequence of the ubiquitous use of DDT. DDT was banned from the market approximately 10 years later.

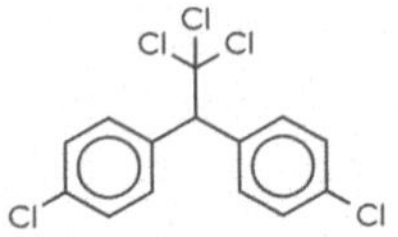

DDT

The toxicity of DDT is due to its interference with sodium channels, leading to the disturbance of normal neurotransmission and causing neurological symptoms. The International Agency for Research on Cancer (IARC) classified DDT in 2015 (monograph 113) as "probably carcinogenic to people, class IIA." The toxicity of DDT was well known in the 1950s, as more than 180 papers on DDT were published from 1950 to 1970. The Stockholm Convention of 2001 on persistent organic pollutants restricts the production and use of Persistant Organis Pollutants (POPs), including DDT. An exception is made for countries where malaria is endemic; in this case, the cautious use of indoor residual spraying with DDT is still permitted (WHO). Replacing DDT with other insecticides is possible, but the alternatives are more expensive (Walker, 2000).

The precautionary principle following the wealth of scientific information in the 1950s was not applied, and it took Carson's book to activate the public, leading to a ban on the use of DDT. The actual situation in which DDT is used in malaria-endemic countries is acceptable.

Bisphenol A (BPA) is used for the production of polycarbonate plastics and epoxy resins. The polycarbonates are used for a.o. plastic bottles and food containers, and the epoxy resins with superior adhesive properties form protective coatings, e.g., inside metal cans, to prevent corrosion by food and drinks. The excellent quality of the products has caused a nearly exponential rise in worldwide production and use of BPA.

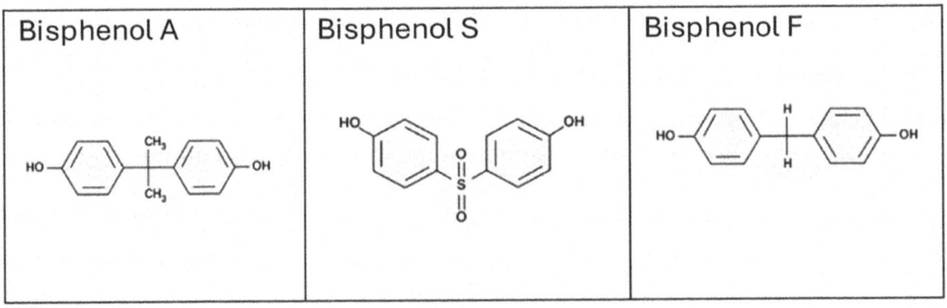

BPA has overt endocrine-disrupting effects (Charkiewicz et al., 2024; Dueñnas-Moreno et al., 2023; Hengstler et al., 2011; Rochester, 2013; Sartain & Hunt, 2016). In 1998, the IARC classified BPA as Class 3 i.e. BPA cannot be classified as a carcinogen to humans, but since then, the carcinogenicity of BPA has been documented extensively (Keri et al., 2007; Prueitt et al., 2023; Ricker et al., 2024; Seachrist et al., 2016). Manufacturers of BPA saw in the 1990s the economic consequences due to human health problems rising and switched from BPA to BPS and some other bisphenols (BPFs), and these products were labeled "BPA-free," suggesting a toxin-free variant of BPA (see the figure for the chemical structure). However, the properties of all bisphenols are very comparable; BPS or BPF are not toxic-free variants. Notably, BPA has been banned from milk bottles for babies, whereas EFSA has reduced the tolerable daily intake from 50 µg/kg/day to 0.2 ng/kg/day, although the feasibility of the latter has been contested (Prueitt & Goodman, 2024). The need for updating the EU guidelines has been expressed (Dueñnas-Moreno et al., 2023).

The balance for BPA is negative, and BPA should have been banned on the basis of precautionary principles. The costs for human health associated with the use of chemicals, among which BPA is a prominent component, are estimated to be 3 billion dollars in the USA (Trasande et al., 2024). The consequences of eliminating BPA are considerable since plastics are embedded in daily life. Changing to plastics that do not need BPA (e.g., polyethylene) would eliminate BPA, but this does not eliminate the occurrence of micro- and nanoplastics (see further). The substitution of plastics for natural products is an option but would require a fundamental change in user mentality, and these products do not have the same flexibility as plastics and are probably more expensive. Here, again, despite the clear negative health impact, the principle was not applied.

Asbestos is a mineral that has been used for many centuries for its non-flammable properties. The mineral exists in at least six different fiber-like forms, each displaying different properties. The crocidolite form has the best fire resistance, although the chrysolite ("white asbestos") form is most commonly used. The commercial use of asbestos started in the 1800s with the first asbestos factory in the USA in 1858. The use of asbestos in the construction business expanded exponentially from the 1900s onward, with peaks during WWII.

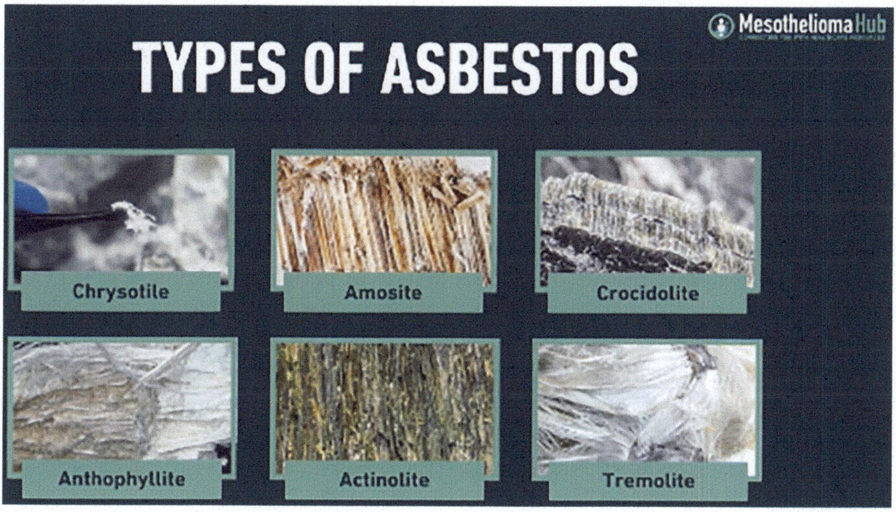

There are multiple medical problems associated with asbestos, but the most striking phenotype is mesothelioma, a nearly uncurable cancerous growth at the level of the lung mesothelium. Mesothelioma and other diseases and the link with asbestos exposure were described from the 1950s leading to severe restrictions on the use and total bans in 1998 in the EU. Notably, the first documented death from mesothelioma dates from 1924, while the description of the presence of asbestos fibers in the lungs of deceased people working in asbestos mines dates from 1906. The epidemiological work of Selikoff in the 1950s, as discussed by McCulloch and Tweedale (2007), the endless discussion on what type of asbestos fiber is the culprit (Tweedale & McCulloch, 2004), and the repulsive attitude toward protection of workers in the UK versus South Africa (McCulloch & Tweedale, 2004) are a small overview of how the safety of workers and by extension of the public was completely disregarded.

The cancerous effect of asbestos exposure is due to the nearly constant ratio of length/width of 3–1 of the fibers, irrespective of their dimensions. When aspirated, the smallest fibers reach the ultimate lower end of the lungs near the diaphragm and cause irritation, inflammation, tissue remodeling, and, ultimately, cancerous growth.

A simple calculation of the benefits and negative effects of asbestos is (again) difficult since the number of people being saved from burning premises using asbestos in construction cannot be estimated. On the negative side, 30,000–40,000 new cases/year of mesothelioma are reported worldwide, with approximately 3000 new cases/year in the EU. This yearly occurrence of mesothelioma is nearly constant, which is the consequence of the long latency and variable period between exposure and disease of between 20 and 40 years. Sufficient alternatives for asbestos, including fiberglass and mineral wool, have been found so that there is no longer a need for its use. Notably, the alternatives are not necessarily without health concerns, particularly concerning the aspiration of the fibers of mineral wool. However, the toxicity is generally lower because the fibers have other dimensions.

The precautionary principle was not applied after the clear demonstration of the negative effects of asbestos in the 1960s. The alternatives were already known, but it was economically much easier to obtain asbestos from mines in Canada, South Africa, and other countries.

So far for the precautionary principle in three clear historical examples, some others have been described by Rosner and Markowitz (2002). How about the precautionary principle in some actual situations?

Per- and polyfluoroalkyl substances or PFAS is the common name for a very large family of products all composed of a carbon chain of variable length substituted with fluor atoms. The major representatives are perfluorooctanoic acid (PFOA) and perfluorooctanesulfonic acid (PFOS), which are synthesized from caprylic acid. The chemicals of the PFAS family are very stable, resist high temperatures and other grim conditions, and are used in firefighting foams. They are known for their oil- and water-repelling properties and are used in very diverse applications.

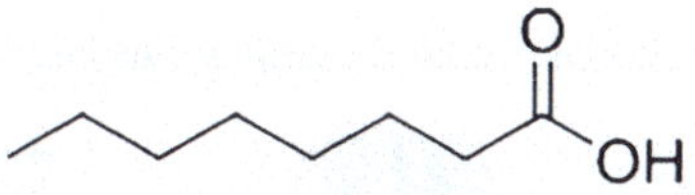

Caprylic acid (fatty acid)

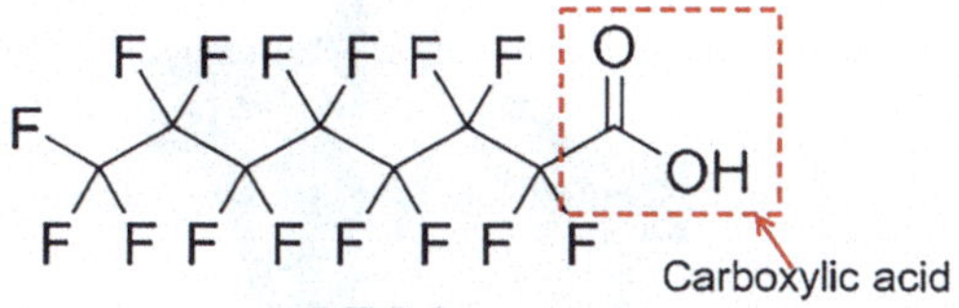

PFOA

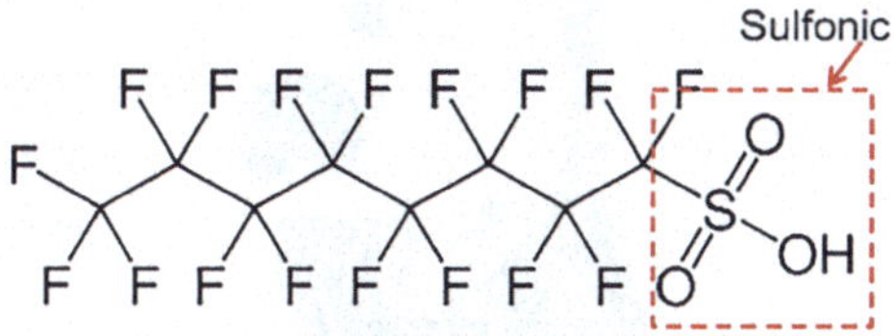

PFOS

PFOAs have been used since 1951 by DuPont for the production of Teflon, which is wrongly considered the nontoxic counterpart of PFAS molecules. Teflon has gained an important place in, e.g., cooking where covering the pans yields a non-sticky and easier way of cooking. Teflon was called the friend of the housewife.

The toxicity profile of PFAS chemicals is highly diverse and includes endocrine-disrupting properties; the IARC coined PFOA as Class I (carcinogenic) and PFOS as Class IIB (possible carcinogenic) to humans. The toxicity of PFASs is high on the agenda of many environmentally friendly organizations and authorities. Restrictions of different kinds are imposed on the manufacturers, but banning PFAS chemicals is not on the table. The focus of manufacturers has now shifted to short-chain molecules (four or six carbon atoms) and TFA (trifluoroacetic acid), which are claimed to be less toxic.

Banning the compounds is urgent, particularly since these “forever chemicals” will remain in the environment for a very long time. The toxicity of members of the PFAS family was well known by a company (3 M) in the 1950s. However, this was not an obstacle to commercializing the compounds while keeping the negative information secret. The films “Dark Waters” (2019) and “The Devil We Know” (2018) have certainly awakened the public and hastened discussions on PFAS. The precautionary principle has deliberately been neglected.

The conclusion for PFAS is straightforward: the production, distribution, and use of PFAS-like molecules should stop ASAP to prevent serious medical problems in the near future. The lack of communication concerning toxic information is mind-boggling.

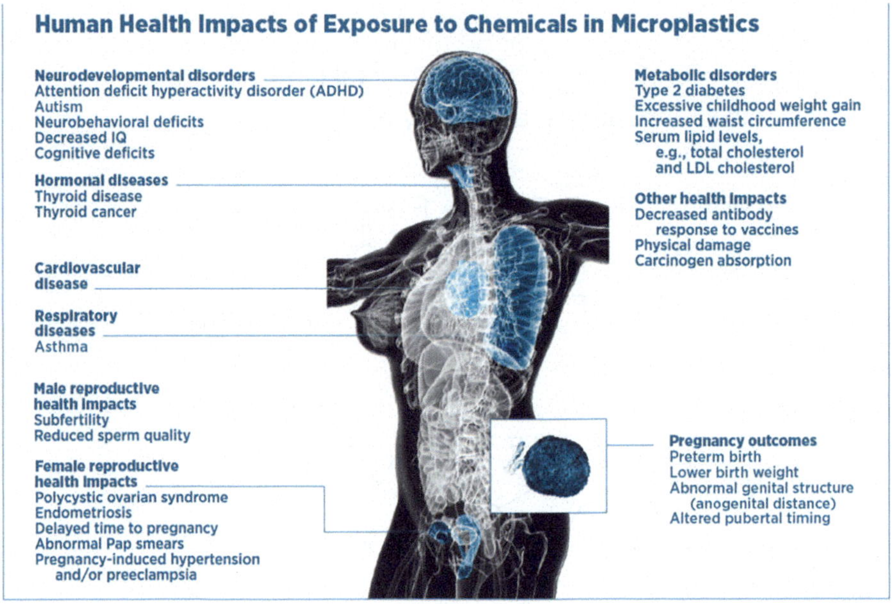

Fig. 2.3 Bodily locations with proven presence of micro- or nanoplastic particles

Micro- and nanoplastics are at the forefront of discussions about environmental pollution (Azfaralariff et al., 2023; Haldar et al., 2023; Hore et al., 2024; Krishnan, 2023; Pelegrini et al., 2023; Vercauteren et al., 2023; Xie et al., 2023; Yarahmadi et al., 2024; Ziani et al., 2023). The presence of plastic particles in various human organs is known (Qiu et al., 2024) such as in seminal fluid (Chen et al., 2024; Li et al., 2024), human blood (Leslie et al., 2022), and the link with human health is obvious (Fontes et al., 2024). Particles are inhaled, which can cause pulmonary problems (Cao et al., 2024). Most importantly, placental translocation has been demonstrated, and plastics are found in umbilical blood and in the brain of fetuses (Medley et al., 2023); all this has been confirmed in animal experiments (Dibbon et al., 2024; Hanrahan et al., 2024; Mercer et al., 2023). Genotoxicity has been ascertained (Cobanoglu et al., 2021; Tagorti & Kaya, 2022). In short, the scientific evidence is overwhelming and sufficient to induce a protective attitude (see Fig. 2.3).

To conclude, with respect to micro- and nanoplastics, the societal impact is immense. Acute toxicity is absent. Long-term toxicity is assumed, although the precise bodily expression of this toxicity is unknown. The Superior Health Council has recently published a piece of advice on micro and nanoplastics: (https://www.health.belgium.be/sites/default/files/uploads/fields/fpshealth_theme_file/hgr_9370_advies_kb_nano.pdf). The absence of biomedical evidence of the toxicity is an argument for manufacturers to question the importance of the results of research. However, if humanity has the ambition to avoid plastic toxicity, interventions are needed. The precautionary principle was not followed at the moment of market introduction.

For *smoking*, we can be short: The evidence for smoking as a causal agent for a number of diseases is overwhelming despite the odds of tobacco-producing companies. The first formally published link between smoking and lung cancer dates from 1956 (Doll & Hill, 1956) and has been repeatedly recognized (Ansari & Jha, 2022; Coleman et al., 2020; Enck et al., 2014; Rahal et al., 2017; Steliga & Dresler, 2011). The replacement of cigarettes with vaping has been proposed as a safe alternative which is commercially interesting but overtly wrong, misleading, and ethically indefensible (Bracken-Clarke et al., 2021).

The cost related to smoking in the USA was 600 billion in 2018 (https://www.cdc.gov/tobacco/data_statistics/fact_sheets/fast_facts/cost-and-expenditures.html), and comparable figures are available for Europe (Koronaiou & Delipalla, 2019). It is estimated that smoking is responsible for 700,000 premature deaths per year, while approximately 25% of the population still smokes (https://health.ec.europa.eu/tobacco/overview_en).

There is no positive effect on human health; in contrast, the costs for society are enormous, and the diseases caused by smoking are devastating. There is not one single reason to continue smoking, and there are many reasons to avoid smoking. A complete ban is indicated.

Glyphosate is the most recent chemical that stands at the forefront of the debate. The further use of the active compound in the EU was permitted in December 2023 for another 10 years until 2033.

Glyphosate is an excellent herbicide that was first synthesized in 1950, patented in 1961 for its metal-binding properties, and marketed in 1974 by the chemical company Monsanto as Roundup. The health problems related to exposure to glyphosate are treated in more detail in the next chapters. Animal experiments suggest the toxic effect of long-term treatment (2 years, daily administration) for a range of tumors in rats and mice, although this is questioned by others. In vitro experiments clearly demonstrated the genotoxic effect on a whole range of cells. The IARC has classified glyphosate as Class 2A (probably carcinogenic to people), whereas other international organizations (e.g., the EPA, EFSA, and ECHA) disagree and claim that glyphosate is not carcinogenic. The endocrine-disrupting properties and major health problems of honey bees have been described, although they are disputed. In terms of human health, acute toxicity is mild and restricted to gastrointestinal disturbances. Long-term toxicity, however, is to be expected; in particular, a link to non-Hodgkin lymphomas has been established among the people most heavily exposed to glyphosate. More recently, the link to Parkinson's disease and other neurological problems has been suggested. A more complete overview of the physical burden of glyphosate exposure has been given recently by the Belgian Superior Health Council (https://www.health.belgium.be/sites/default/files/uploads/fields/fpshealth_theme_file/20200303_hgr-9561_glyphosate_vweb.pdf).

The consequences of banning glyphosate would be significant. Glyphosate is used mainly on crops needed to feed cattle, which is needed to feed meat-eating people. Hence, banning glyphosate would imply a major shift in eating culture, particularly in the Western world. It is unlikely that this will readily happen despite the advice from the EAT-Lancet Commission on healthy diets of 2019 (https://www.thelancet.com/journals/lancet/article/PIIS0140-6736(18)31788-4/abstract). Whether the precautionary principle has been discussed during the recent glyphosate evaluation process is not known. The term "precautionary principle" was not mentioned once in the EFSA peer review document on the permit to use glyphosate till 2033 (https://www.efsa.europa.eu/en/efsajournal/pub/8164).

In the foregoing chapters, seven items were discussed: three simple chemicals (DDT, BPA, and glyphosate), one complex undefined mixture (micro- and nanoplastics), one family of chemicals (PFAS family), one mineral (asbestos), and one bad habit (smoking). This choice was not haphazardly but was based on (1) their ubiquitous use and consequent universal human exposure, (2) intense legal and regulatory efforts in recent decades based on science but subject to external forces, and (3) high financial implications both during manufacturing, production, and introduction at the market and upon potential bans on important societal implications such as employment in the chemical industry. There is little acute toxicity, but all information points in the direction of health problems in the middle or long term. The fact that the bodily expression of this toxicity is not fully known or concealed at the moment of its introduction on the market or during the re-evaluation process, is a good reason to consider the application of the precautionary principle. The conclusion is that the precautionary principle is not applied where it should be. What are possible actions to solve the riddle?

Since any decision depends on the quality of the risk assessment, better toxicological data are needed.

First, the study of the risks for human health due to exposure to a chemical must be performed by completely independent scientists and organizations, according to the highest GLP standards, and the relative weight of all available tests in the decision algorithm must be taken into account in the decision algorithm. The future of toxicity testing has been discussed before (Andersen et al., 2010; Krewski et al., 2020) and, more specifically, for future toxicity testing of pesticides (Reffstrup et al., 2010; Seiber et al., 2014).

Second, all tests and results must be open to peer review taking confidentiality into consideration. The interpretation and conclusions of the tests are performed by an independent body, not related to the companies, not related to the organization that performed the experiments, and independent of the political authorities.

Third, all the people and organizations involved must agree beforehand with the planned experiments and the planned procedure for evaluating the results of the tests. In particular, all have to agree on the relative weights of the tests in the final decision.

Fourth, the ultimate conclusion can still apply—if necessary—the precautionary principle. Whatever the conclusion and decision is, it will be transparent.

I am aware that all this has been said before and that there is a minimal chance that the above is realizable. Companies will probably never agree on this and put the lobby companies to work. Political authorities are likely to consider financial investments too high.

In conclusion, as a scientists, writing a chapter on the precautionary principle is highly confusing. On the one hand, scientific information on the subject is available, although, by definition, it is timely, incomplete, and debatable. The final very uneasy feeling is that the precautionary principle is considered or disregarded haphazardly, with economic considerations such as primality, not human health.

Taking this into account, false positive (too restrictive) and false negative (too loose) applications of the precautionary principle exist and must be avoided. This leads to the question of how much evidence you need to make the right conclusions.

All this is accompanied by important philosophical-ethical questions, such as whether we need nanoplastics to scrub our skin, a non-sticky pan to fry the meat, or water-repellent bathing suits for surfers. It seems that the artificially created needs have evolved into an allegedly vital and indispensable benefit of daily life. Degrowth or even reconsideration of the principle of the necessity of economic growth, which is the main drive for introducing chemicals, is currently nearly impossible (Aagaard et al., 2013).

The conclusion is that in the given examples, historical and actual, the precautionary principle has NOT been followed correctly, but other priorities prevailed, the most important being financial opportunities. This will have important evolutionary consequences (Bolognesi et al., 2022) or, at extreme levels, jeopardize the human race (https://iris.who.int/handle/10665/46971).

References

Aagaard, K., Petrosino, J., Keitel, W., Watson, M., Katancik, J., Garcia, N., et al. (2013). The Human Microbiome Project strategy for comprehensive sampling of the human microbiome and why it matters. *FASEB Journal, 27*(3), 1012–1022. https://doi.org/10.1096/fj.12-220806

Andersen, M. E., Al-Zoughool, M., Croteau, M., Westphal, M., & Krewski, D. (2010). The future of toxicity testing. *Journal of Toxicology and Environmental Health-Part B-Critical Reviews, 13*(2–4), 163–196. https://doi.org/10.1080/10937404.2010.483933

Ansari, K. K., & Jha, A. (2022). Causes of cancer in the world: Comparative risk assessment of nine behavioral and environmental risk factors. *Cureus Journal of Medical Science, 14*(9), 14. https://doi.org/10.7759/cureus.28875

Azfaralariff, A., Lazim, A. M., Amran, N. H., Mukhtar, N. H., Bakri, N. D., Azrihan, N. N., & Mohamad, M. (2023). Mini review of microplastic pollutions and its impact on the environment and human health. *Waste Management & Research, 8*, 1219. https://doi.org/10.1177/0734242x231155395

Beilmann, M., Boonen, H., Czich, A., Dear, G., Hewitt, P., Mow, L., et al. (2019). Optimizing drug discovery by investigative toxicology: Current and future trends. *Altex-Alternatives to Animal Experimentation, 36*(2), 289–313. https://doi.org/10.14573/altex.1808181

Benigni, R., & Bossa, C. (2011). Alternative strategies for carcinogenicity assessment: An efficient and simplified approach based on in vitro mutagenicity and cell transformation assays. *Mutagenesis, 26*(3), 455–460. https://doi.org/10.1093/mutage/ger004

Benigni, R., Bossa, C., Battistelli, C. L., & Tcheremenskaia, O. (2013). IARC Classes 1 and 2 carcinogens are successfully identified by an alternative strategy that detects DNA-reactivity and cell transformation ability of chemicals. *Mutation Research-Genetic Toxicology and Environmental Mutagenesis, 758*(1–2), 56–61. https://doi.org/10.1016/j.mrgentox.2013.09.006

Benigni, R., Battistelli, C. L., Bossa, C., Giuliani, A., & Tcheremenskaia, O. (2015). Alternative toxicity testing: Analyses on skin sensitization, ToxCast phases I and II, and carcinogenicity provide indications on how to model mechanisms linked to adverse outcome pathways. *Journal of Environmental Science and Health Part C-Environmental Carcinogenesis & Ecotoxicology Reviews, 33*(4), 422–443. https://doi.org/10.1080/10590501.2015.1096885

Bolognesi, G., Bacalini, M. G., Pirazzini, C., Garagnani, P., & Giuliani, C. (2022). Evolutionary implications of environmental toxicant exposure. *Biomedicines, 10*(12), 26. https://doi.org/10.3390/biomedicines10123090

Bracken-Clarke, D., Kapoor, D., Baird, A. M., Buchanan, P. J., Gately, K., Cuffe, S., & Finn, S. P. (2021). Vaping and lung cancer – A review of current data and recommendations. *Lung Cancer, 153*, 11–20. https://doi.org/10.1016/j.lungcan.2020.12.030

Cao, Y., Zhao, Q., Jiang, F. S., Geng, Y. X., Zhang, L. F., Li, C., et al. (2024). Interactions between inhalable aged microplastics and lung surfactant: Potential pulmonary health risks. *Environmental Research, 245*, 10. https://doi.org/10.1016/j.envres.2023.117803

Charkiewicz, A. E., Omeljaniuk, W. J., & Niklinski, J. (2024). Bisphenol A-what do we know? A global or local approach at the public health risk level. *International Journal of Molecular Sciences, 25*(11), 19. https://doi.org/10.3390/ijms25116229

Chen, Y., Cheng, C., Xu, W. Q., Cui, Y. F., Tian, Y., Jiang, Y. L., et al. (2024). Occurrence, toxicity and removal of polystyrene microplastics and nanoplastics in human sperm. *Environmental Chemistry Letters, 7*, 2159. https://doi.org/10.1007/s10311-024-01752-0

Cobanoglu, H., Belivermis, M., Sikdokur, E., Kilic, O., & Cayir, A. (2021). Genotoxic and cytotoxic effects of polyethylene microplastics on human peripheral blood lymphocytes. *Chemosphere, 272*, 6. https://doi.org/10.1016/j.chemosphere.2021.129805

Coleman, N. C., Burnett, R. T., Higbee, J. D., Lefler, J. S., Merrill, R. M., Ezzati, M., et al. (2020). Cancer mortality risk, fine particulate air pollution, and smoking in a large, representative cohort of US adults. *Cancer Causes & Control, 31*(8), 767–776. https://doi.org/10.1007/s10552-020-01317-w

Cox, L. A. (2018). Modernizing the Bradford Hill criteria for assessing causal relationships in observational data. *Critical Reviews in Toxicology, 48*(8), 682–712. https://doi.org/10.1080/10408444.2018.1518404

de Almeida, N., Castiel, L. D., & Ayres, J. R. (2009). Risk: Basic concept of epidemiology. *Salud Colectiva, 5*(3), 323–344. https://doi.org/10.1590/s1851-82652009000300003

De Troeyer, K., Casas, L., Bijnens, E. M., Bruckers, L., Covaci, A., De Henauw, S., et al. (2022). Higher proportion of agricultural land use around the residence is associated with higher urinary concentrations of AMPA, a glyphosate metabolite. *International Journal of Hygiene and Environmental Health, 246*, 8. https://doi.org/10.1016/j.ijheh.2022.114039

Dibbon, K. C., Mercer, G., Maekawa, A. S., Hanrahan, J., Steeves, K. L., Ringer, L. C. M., et al. (2024). Polystyrene micro- and nanoplastics cause placental dysfunction in mice. *Biology of Reproduction, 110*(1), 211–218. https://doi.org/10.1093/biolre/ioad126

Doll, R., & Hill, A. B. (1956). Lung cancer and other causes of death in relation to smoking – A 2nd report on the mortality of British doctors. *British Medical Journal, 2*(NOV10), 1071–1081. https://doi.org/10.1136/bmj.2.5001.1071

Dueñnas-Moreno, J., Mora, A., Kumar, M., Meng, X. Z., & Mahlknecht, J. (2023). Worldwide risk assessment of phthalates and bisphenol A in humans: The need for updating guidelines. *Environment International, 181*, 13. https://doi.org/10.1016/j.envint.2023.108294

Enck, P., Weimer, K., & Klosterhalfen, S. (2014). Are all placebo respondents non-smokers? *Medical Hypotheses, 83*(3), 355–358. https://doi.org/10.1016/j.mehy.2014.06.012

Fontes, B. L. M., Souza, L., de Oliveira, A., da Fonseca, R. N., Neto, M. P. C., & Pinheiro, C. R. (2024). The possible impacts of nano and microplastics on human health: Lessons from experimental models across multiple organs. *Journal of Toxicology and Environmental Health-Part B-Critical Reviews, 27*(4), 153–187. https://doi.org/10.1080/10937404.2024.2330962

Goldstein, B. D., & Carruth, R. S. (2005). Implications of the precautionary principle: Is it a threat to science? (Reprinted from the European Journal of Oncology Library, vol 2, 2003). *Human and Ecological Risk Assessment, 11*(1), 209–219. https://doi.org/10.1080/10807030590920033

Gross, S. A., & Fedak, K. M. (2015). Applying a weight-of-evidence approach to evaluate relevance of molecular landscapes in the exposure-disease paradigm. *Biomed Research International, 2015*, 11. https://doi.org/10.1155/2015/515798

Haldar, S., Muralidaran, Y., Miguez, D., Mulla, S. I., & Mishra, P. (2023). Eco-toxicity of nanoplastics and its implication on human metabolism: Current and future perspective. *Science of the Total Environment, 861*, 12. https://doi.org/10.1016/j.scitotenv.2022.160571

Hanrahan, J., Steeves, K. L., Locke, D. P., O'Brien, T. M., Maekawa, A. S., Amiri, R., et al. (2024). Maternal exposure to polyethylene micro- and nanoplastics impairs umbilical blood flow but not fetal growth in pregnant mice. *Scientific Reports, 14*(1), 7. https://doi.org/10.1038/s41598-023-50781-2

Hartung, T. (2019a). AI Versus Animal Testing. *Scientist, 33*(5), 23–23. Retrieved from <Go to ISI>://WOS:000466832500008.

Hartung, T. (2019b). Predicting toxicity of chemicals: Software beats animal testing. *EFSA Journal, 17*, 8. https://doi.org/10.2903/j.efsa.2019.e170710

Hemphill, T. A. (2020). The innovation governance dilemma: Alternatives to the precautionary principle. *Technology in Society, 63*, 9. https://doi.org/10.1016/j.techsoc.2020.101381

Hengstler, J. G., Foth, H., Gebel, T., Kramer, P. J., Lilienblum, W., Schweinfurth, H., et al. (2011). Critical evaluation of key evidence on the human health hazards of exposure to bisphenol A. *Critical Reviews in Toxicology, 41*(4), 263–291. https://doi.org/10.3109/10408444.2011.558487

Hill, A. B. (1965). Environment and disease – Association or causation. *Proceedings of the Royal Society of Medicine-London, 58*(5), 295-+. https://doi.org/10.1177/003591576505800503

Högberg, J., & Järnberg, J. (2023). Approaches for the setting of occupational exposure limits (OELs) for carcinogens. *Critical Reviews in Toxicology, 53*(3), 131–167. https://doi.org/10.1080/10408444.2023.2218887

Hore, M., Bhattacharyya, S., Roy, S., Sarkar, D., & Biswas, J. K. (2024). Human exposure to dietary microplastics and health risk: A comprehensive review. *Reviews of Environmental Contamination and Toxicology, 262*(1), 26. https://doi.org/10.1007/s44169-024-00066-0

Keri, R. A., Ho, S. M., Hunt, P. A., Knudsen, K. E., Soto, A. M., & Prins, G. S. (2007). An evaluation of evidence for the carcinogenic activity of bisphenol A. *Reproductive Toxicology, 24*(2), 240–252. https://doi.org/10.1016/j.reprotox.2007.06.008

Knight, A., Bailey, J., & Balcombe, J. (2006a). Animal carcinogenicity studies: 1. Poor human predictivity. *Atla-Alternatives to Laboratory Animals, 34*(1), 19–27. https://doi.org/10.1177/026119290603400117

Knight, A., Bailey, J., & Balcombe, J. (2006b). Animal carcinogenicity studies: 2. Obstacles to extrapolation of data to humans. *Atla-Alternatives to Laboratory Animals, 34*(1), 29–38. https://doi.org/10.1177/026119290603400118

Knight, A., Bailey, J., & Balcombe, J. (2006c). Animal carcinogenicity studies: 3. Alternatives to the bioassay. *Atla-Alternatives to Laboratory Animals, 34*(1), 39–48. https://doi.org/10.1177/026119290603400119

Koronaiou, K., & Delipalla, S. (2019). The economic cost of tobacco smoking and secondhand smoke in Greece: Musculoskeletal disorders the leading contributor to smoking related morbidity. *Tobacco Prevention & Cessation, 5*, 10. https://doi.org/10.18332/tpc/113091

Krewski, D., Andersen, M. E., Mantus, E., & Zeise, L. (2009). Toxicity testing in the 21st century: Implications for human health risk assessment. *Risk Analysis, 29*(4), 474–479. https://doi.org/10.1111/j.1539-6924.2008.01150.x

Krewski, D., Andersen, M. E., Tyshenko, M. G., Krishnan, K., Hartung, T., Boekelheide, K., et al. (2020). Toxicity testing in the 21st century: Progress in the past decade and future perspectives. *Archives of Toxicology, 94*(1), 1–58. https://doi.org/10.1007/s00204-019-02613-4

Krishnan, K. (2023). A systematic review on the impact of micro-nanoplastics exposure on human health and diseases. *Biointerface Research in Applied Chemistry, 13*(4), 12. https://doi.org/10.33263/briac134.381

Ku, R. H. (2001). Practical aspects of setting occupational exposure limits for pharmaceuticals. *Abstracts of Papers of the American Chemical Society, 221*, U275–U275. Retrieved from <Go to ISI>://WOS:000168824702110.

Landis, W. G. (2012). Comment on Fox et al. (2012): What to do with NOECs/NOELs. *Integrated Environmental Assessment and Management, 8*(4), 766–767. https://doi.org/10.1002/ieam.1347

Lash, T. L., Ahern, T. P., Collin, L. J., Fox, M. P., & MacLehose, R. F. (2021). Bias analysis gone bad. *American Journal of Epidemiology, 190*(8), 1604–1612. https://doi.org/10.1093/aje/kwab072

Leslie, H. A., van Velzen, M. J. M., Brandsma, S. H., Vethaak, A. D., Garcia-Vallejo, J. J., & Lamoree, M. H. (2022). Discovery and quantification of plastic particle pollution in human blood. *Environment International, 163*, 8. https://doi.org/10.1016/j.envint.2022.107199

Li, N., Yang, H. J., Dong, Y. L., Wei, B., Liang, L. Y., Yun, X., et al. (2024). Prevalence and implications of microplastic contaminants in general human seminal fluid: A Raman spectroscopic study. *Science of the Total Environment, 937*, 7. https://doi.org/10.1016/j.scitotenv.2024.173522

Maertens, A., Golden, E., Luechtefeld, T. H., Hoffmann, S., Tsaioun, K., & Hartung, T. (2022). Probabilistic risk assessment – The keystone for the future of toxicology. *Altex-Alternatives to Animal Experimentation, 39*(1), 3–29. https://doi.org/10.14573/altex.2201081

Maertens, A., Antignac, E., Benfenati, E., Bloch, D., Fritsche, E., Hoffmann, S., et al. (2024a). The probable future of toxicology – Probabilistic risk assessment. *Altex-Alternatives to Animal Experimentation, 41*(2), 273–281. https://doi.org/10.14573/altex.2310301

Maertens, A., Luechtefeld, T., Knight, J., & Hartung, T. (2024b). Alternative methods go green! Green toxicology as a sustainable approach for assessing chemical safety and designing safer chemicals. *Altex-Alternatives to Animal Experimentation, 41*(1), 3–19. Retrieved from <Go to ISI>://WOS:001147333200004.

McCulloch, J., & Tweedale, G. (2004). Double standards: The multinational asbestos industry and asbestos-related disease in South Africa. *International Journal of Health Services, 34*(4), 663–679. https://doi.org/10.2190/f06d-le93-m6lh-xj0b

McCulloch, J., & Tweedale, G. (2007). Shooting the messenger: The vilification of Irving. J. Selikoff. *International Journal of Health Services, 37*(4), 619–634. https://doi.org/10.2190/HS.37.4.b

McHenry, L. B. (2018). The Monsanto Papers: Poisoning the scientific well. *International Journal of Risk & Safety in Medicine, 29*(3–4), 193–205. https://doi.org/10.3233/jrs-180028

Medley, E. A., Spratlen, M. J., Yan, B. Z., Herbstman, J. B., & Deyssenroth, M. A. (2023). A systematic review of the placental translocation of micro- and nanoplastics. *Current Environmental Health Reports, 13*, 99. https://doi.org/10.1007/s40572-023-00391-x

Mercer, G. V., Harvey, N. E., Steeves, K. L., Schneider, C. M., Sled, J. G., Macgowan, C. K., et al. (2023). Maternal exposure to polystyrene nanoplastics alters fetal brain metabolism in mice. *Metabolomics, 19*(12), 8. https://doi.org/10.1007/s11306-023-02061-3

Münzel, T., Sorensen, M., Lelieveld, J., Hahad, O., Al-Kindi, S., Nieuwenhuijsen, M., et al. (2021). Heart healthy cities: Genetics loads the gun but the environment pulls the trigger. *European Heart Journal, 42*(25), 2422–2438. https://doi.org/10.1093/eurheartj/ehab235

Nohmi, T. (2018). Thresholds of genotoxic and non-genotoxic carcinogens. *Toxicological Research, 34*(4), 281–290. https://doi.org/10.5487/TR.2018.34.4.281

Pelegrini, K., Pereira, T. C. B., Maraschin, T. G., Teodoro, L. D., Basso, N. R. D., Galland, G., et al. (2023). Micro- and nanoplastic toxicity: A review on size, type, source, and test-organism implications. *Science of the Total Environment, 878*, 14. https://doi.org/10.1016/j.scitotenv.2023.162954

Prueitt, R. L., & Goodman, J. E. (2024). Evidence evaluated by European Food Safety Authority does not support lowering the temporary tolerable daily intake for bisphenol A. *Toxicological Sciences, 198*(2), 185–190. https://doi.org/10.1093/toxsci/kfad136

Prueitt, R. L., Hixon, M. L., Fan, T. Y., Olgun, N. S., Piatos, P., Zhou, J., & Goodman, J. E. (2023). Systematic review of the potential carcinogenicity of bisphenol A in humans. *Regulatory Toxicology and Pharmacology, 142*, 38. https://doi.org/10.1016/j.yrtph.2023.105414

Qiu, H., Li, J. J., Chen, G. Q., Cao, T. Y., Li, J., Li, X., & Qiu, R. L. (2024). Penetration of micro/nanoplastics into biological barriers in organisms and associated health effects. *Chinese Science Bulletin-Chinese, 69*(6), 717–730. https://doi.org/10.1360/tb-2023-0295

Rahal, Z., El Nemr, S., Sinjab, A., Chami, H., Tfayli, A., & Kadara, H. (2017). Smoking and lung cancer: A geo-regional perspective. *Frontiers in Oncology, 7*, 7. https://doi.org/10.3389/fonc.2017.00194

Reffstrup, T. K., Larsen, J. C., & Meyer, O. (2010). Risk assessment of mixtures of pesticides. Current approaches and future strategies. *Regulatory Toxicology and Pharmacology, 56*(2), 174–192. https://doi.org/10.1016/j.yrtph.2009.09.013

Ricker, K., Cheng, V. E. S., Hsieh, C. J., Tsai, F. C., Osborne, G., Li, K. T., et al. (2024). Application of the key characteristics of carcinogens to bisphenol A. *International Journal of Toxicology, 43*(3), 253–290. https://doi.org/10.1177/10915818231225161

Rochester, J. R. (2013). Bisphenol A and human health: A review of the literature. *Reproductive Toxicology, 42*, 132–155. https://doi.org/10.1016/j.reprotox.2013.08.008

Rosner, D., & Markowitz, G. (2002). Industry challenges to the principle of prevention in public health: The precautionary principle in historical perspective. *Public Health Reports, 117*(6), 501–512. https://doi.org/10.1016/s0033-3549(04)50195-4

Russo, D. P., Strickland, J., Karmaus, A. L., Wang, W. Y., Shende, S., Hartung, T., et al. (2019). Nonanimal models for acute toxicity evaluations: Applying data-driven profiling and read-across. *Environmental Health Perspectives, 127*(4), 14. https://doi.org/10.1289/ehp3614

Sartain, C. V., & Hunt, P. A. (2016). An old culprit but a new story: bisphenol A and "NextGen" bisphenols. *Fertility and Sterility, 106*(4), 820–826. https://doi.org/10.1016/j.fertnstert.2016.07.1114

Seachrist, D. D., Bonk, K. W., Ho, S. M., Prins, G. S., Soto, A. M., & Keri, R. A. (2016). A review of the carcinogenic potential of bisphenol A. *Reproductive Toxicology, 59*, 167–182. https://doi.org/10.1016/j.reprotox.2015.09.006

Seiber, J. N., Coats, J., Duke, S. O., & Gross, A. D. (2014). Biopesticides: State of the art and future opportunities. *Journal of Agricultural and Food Chemistry, 62*(48), 11613–11619. https://doi.org/10.1021/jf504252n

Shimonovich, M., Pearce, A., Thomson, H., Keyes, K., & Katikireddi, S. V. (2021). Assessing causality in epidemiology: Revisiting Bradford Hill to incorporate developments in causal thinking. *European Journal of Epidemiology, 36*(9), 873–887. https://doi.org/10.1007/s10654-020-00703-7

Steliga, M. A., & Dresler, C. M. (2011). Epidemiology of lung cancer: Smoking, secondhand smoke, and genetics. *Surgical Oncology Clinics of North America, 20*(4), 605-+. https://doi.org/10.1016/j.soc.2011.07.003

Tagorti, G., & Kaya, B. (2022). Genotoxic effect of microplastics and COVID-19: The hidden threat. *Chemosphere, 286*, 14. https://doi.org/10.1016/j.chemosphere.2021.131898

Trasande, L., Krithivasan, R., Park, K., Obsekov, V., & Belliveau, M. (2024). Chemicals used in plastic materials: An estimate of the attributable disease burden and costs in the United States. *Journal of the Endocrine Society, 8*(2), 9. https://doi.org/10.1210/jendso/bvad163

Tweedale, G., & McCulloch, J. (2004). Chrysophiles versus chrysophobes – The white asbestos controversy, 1950s–2004. *Isis, 95*(2), 239–259. https://doi.org/10.1086/426196

Ucar, T., & Hall, F. R. (2001). Windbreaks as a pesticide drift mitigation strategy: A review. *Pest Management Science, 57*(8), 663–675. https://doi.org/10.1002/ps.341

Vercauteren, M., Semmouri, I., Van Acker, E., Pequeur, E., Van Esch, L., Uljee, I., et al. (2023). Assessment of road run-off and domestic wastewater contribution to microplastic pollution in a densely populated area (Flanders, Belgium)*. *Environmental Pollution, 333*, 9. https://doi.org/10.1016/j.envpol.2023.122090

Virolainen, S. J., VonHandorf, A., Viel, K., Weirauch, M. T., & Kottyan, L. C. (2023). Gene-environment interactions and their impact on human health. *Genes and Immunity, 24*(1), 1–11. https://doi.org/10.1038/s41435-022-00192-6

Walker, K. (2000). Cost-comparison of DDT and alternative insecticides for malaria control. *Medical and Veterinary Entomology, 14*(4), 345–354. https://doi.org/10.1046/j.1365-2915.2000.00262.x

Wang, Z. Y., Walker, G. W., Muir, D. C. G., & Nagatani-Yoshida, K. (2020). Toward a global understanding of chemical pollution: A first comprehensive analysis of national and regional chemical inventories. *Environmental Science & Technology, 54*(5), 2575–2584. https://doi.org/10.1021/acs.est.9b06379

Weed, D. L. (2018). Analogy in causal inference: Rethinking Austin Bradford Hill's neglected consideration. *Annals of Epidemiology, 28*(5), 343–346. https://doi.org/10.1016/j.annepidem.2018.03.004

Xie, H. X., Tian, X., Lin, X. Y., Chen, R., Hameed, S., Wang, L. M., et al. (2023). Nanoplastic-induced biological effects in vivo and in vitro: An overview. *Reviews of Environmental Contamination and Toxicology, 261*(1), 26. https://doi.org/10.1007/s44169-023-00027-z

Yarahmadi, A., Heidari, S., Sepahvand, P., Afkhami, H., & Kheradjoo, H. (2024). Microplastics and environmental effects: Investigating the effects of microplastics on aquatic habitats and their impact on human health. *Frontiers in Public Health, 12*, 22. https://doi.org/10.3389/fpubh.2024.1411389

Zhang, Q. A., Bhattacharya, S., Andersen, M. E., & Conolly, R. B. (2010). Computational systems biology and dose-response modeling in relation to new directions in toxicity testing. *Journal of Toxicology and Environmental Health-Part B-Critical Reviews, 13*(2–4), 253–276. https://doi.org/10.1080/10937404.2010.483943

Ziani, K., Ionita-Mindrican, C. B., Mititelu, M., Neacsu, S. M., Negrei, C., Morosan, E., et al. (2023). Microplastics: A real global threat for environment and food safety: A state of the art review. *Nutrients, 15*(3), 34. https://doi.org/10.3390/nu15030617

Chapter 3
Glyphosate and Cancer

After the publication in 2015 by the IARC that glyphosate is probably a carcinogen for humans, the carcinogenicity of glyphosate has been at the center of nearly all discussions on glyphosate. What follows are a short introductory note on carcinogenesis, the points of view of the world's most important organizations, and the literature. Finally, based on this information, a suggestion on the carcinogenicity of glyphosate is given.

Introduction on Carcinogenesis

For a detailed recent review of carcinogenesis, please consult the NIH (https://www.ncbi.nlm.nih.gov/books/NBK604463/).

Carcinogenesis is a complex biological phenomenon in which the control of the normal growth of healthy cells within a particular tissue is lost. "Normal growth" means the equilibrium between cells dying by programmed cell death, called apoptosis, and new cells replacing the dead cells, called proliferation, that is under strict enzymatic control and specific circumstances proper to the tissue. The result is a healthy, fully active organ limited in volume. When this control fails, abnormal growth can occur. This growth is called benign when no surrounding tissues are affected or malignant when the growing lump of tissue affects other tissues (invasion) or can grow secondary tumors in other organs (metastasis).

The steps of carcinogenesis include (1) the exposure of normal cells to a carcinogenic substance. (2) This exposure results in alterations in the cell's genetic makeup at the level of the DNA. (3) If the conditions are favorable, this mutated cell can develop into a tumor, including further abnormalities in the DNA.

The first step, "initiation," refers to the interaction of a substance with a healthy cell. This substance can be a foreign chemical (e.g., benzene), some pharmacological compounds (e.g., some drugs used in cancer therapy), a group of chemicals (e.g.,

N. Fraeyman, *Glyphosate 2023–2033*,
https://doi.org/10.1007/978-3-031-97540-0_3

smoking), a virus (e.g., the human papillomavirus), UV radiation (e.g., excessive exposure to the sun), or ionic radiation (e.g., X-rays during medical exams). Oncogenes, which are present in healthy cells and whose activity is depressed, might become active because of a defect in the controlling system. Alternatively, the tumor suppressor gene(s) that prevent the development of cancer might become defective due to mutations. The best-known tumor suppressor gene is described in the literature as the "p53 gene." Some other substances proper to the body, such as those present after chronic infections, can also lead to cancerous growth.

Some of these molecules act directly on the DNA through a chemical reaction, distorting the well-known double helix of the DNA and hampering its multiplication during the cell division process. This is the case with, e.g., the group of alkylating drugs such as cyclophosphamide. Some other molecules do not act directly on the DNA but rather on other biological systems or molecules closely linked to the DNA. This is called the epigenetic effect, which might also result in cancerous growth. Other molecules do not cause mutations on their own but facilitate the cancerous process by encouraging the growth of already mutated cells. The characteristics of these tumor inducers are very diverse and include, e.g., phorbol esters, certain hormones, and cytokines.

The second step, "progression," means that the circumstances are favorable for the survival and further outgrowth of the cancerous cell. The whole bodily constitution contributes in some way: Hormones or chemicals from infectious tissue might stimulate growth, whereas others, such as the immune system, might inhibit it. The tumor can produce substances that promote the growth of veins in the tumor to provide the tumor cells with nutrients.

The third step, "progression," means that the cells become very deregulated, and many other mutations occur, leading to further deviation from normal control systems. Cells acquire new properties, such as the capacity to invade other organs or to spread into the body and form secondary tumors or metastasize to other organs.

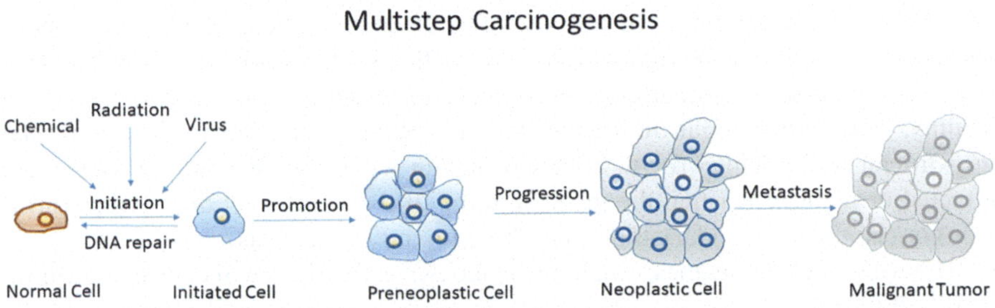

The prevalence of cancer is, on average, approximately 30%, despite universal exposure to carcinogens such as sunlight and natural radiation. This figure is much lower in the younger population and higher in the older population. The reason is that one single mutation following one interaction of a chemical with the DNA is not sufficient to induce cancerous growth. It is assumed that between 4 and 7 critical mutations are needed to transform a normal cell into a cancerous cell. If the DNA change is heritable, the damage is transferred to the next generations of cells, and during each generation, DNA damage might occur. All this takes time, which is the reason why older people are more prone to cancer than younger people. However, this is not always the case. Some cancers (e.g., pancreatic cancer and small cell lung cancer) may grow quickly.

Some cell types are more susceptible to carcinogenesis than others because of the higher cell division rate and because the genetic material of cells is most vulnerable during the cell division process since, at that moment in the cell's life cycle, the DNA strains separate and reanneal with new nucleotides to form new cells. Neuronal cells divide very slowly and have very stable DNA during their life span. Brain tumors are mostly tumors of glial cells, i.e., neuron-supporting cells, that divide continuously.

The second reason for the low prevalence of cancers is that normal human cells possess very efficient repair systems that detect and restore errors in the DNA machinery. These repair systems, involving several enzymes that work in close and fine-tuned collaboration, are needed to correct for the numerous spontaneous, naturally occurring mutations in each cell every day. Most of these events are corrected and remain without consequences. However, if cells are hit at the level of the genes controlling the repair system, then repair is not evident, and more mutations will occur and gain viability, leading to an increased risk for cancerous growth (see above, the p53 suppressor gene).

Taken together, carcinogenesis is a complex process ultimately leading to an event at the level of the genetic material of the cell. If all circumstances are fulfilled, this event can lead to the growth of a tumor.

Carcinogenicity of Glyphosate

In what follows is the review of the point of view of four organizations: the EPA in the USA, the EFSA and ECHA in Europe, and the IARC for the World Health Organization (WHO) together with older and recent literature trying to answer the following questions: (1) What are the points of view of these organizations, and how are they expressed in scientific terms? (2) What method was used to reach this stand, and what information was used in support? Part of this text was published under the HGR advice of 2020 (https://www.health.belgium.be/sites/default/files/uploads/fields/fpshealth_theme_file/20200303_hgr-9561_glyphosate_vweb.pdf).

The *United States Environmental Protection Agency (EPA)* classified glyphosate in 1985 as a Class C carcinogen: "suggestive evidence of carcinogenic potential with limited animal evidence and little or no human data." In 1991, the standpoint of the EPA became "evidence of non-carcinogenicity for humans," with classification as a Group E carcinogen. The EPA's 1991 report can be found at https://archive.epa.gov/pesticides/chemicalsearch/chemical/foia/web/pdf/103601/417300-1991-10-30a.pdf.

Glyphosate was re-reviewed for carcinogenicity in 2017, and the EPA concluded that "glyphosate is unlikely to be carcinogenic to humans when the product is used according to the pesticide label." Interestingly, the EPA distinguishes between the potential of carcinogenicity in biomedical terms and the risk to humans, depending on the extent of exposure.

A detailed overview of the method used by the EPA is given in https://www.epa.gov/fera/risk-assessment-carcinogenic-effects, https://www.greenfacts.org/glossary/def/epa-cancer-classification.htm, and https://www.epa.gov/pesticide-science-and-assessing-pesticide-risks/evaluating-pesticides-carcinogenic-potential#a.

The EPA uses human data, data from long-term experimental animal bioassays, and a variety of short-term tests, such as genetic toxicity testing, pharmacokinetic and pharmacodynamic data, metabolic studies, and studies of structure–activity relationships. Any data, peer-reviewed or not, may be used to support the organization's point of view, provided it is of sufficient scientific quality according to EPA-determined criteria.

The EPA relies heavily on unpublished information, mostly delivered by the industry during the regulatory process. It is tempting to speculate that these studies are biased because they originate from a company with economic interests in the outcome. However, such studies are mostly carried out under Good Laboratory Practices (GLPs), the aim of which is to guarantee the correctness of the experimental results. While the raw data are probably trustworthy, both the conclusions of these studies and speculations on their nature are subject to debate.

Many other US organizations have opinions on the carcinogenicity of glyphosate, including the National Institutes of Health's Agricultural Health Survey, the National Toxicology Program, and the Council on Environmental Quality. The positions of these bodies are not taken into consideration.

To comment and conclude on the EPA, the EPA has gained much credibility and trust as a governmental organization performing experiments in a large network of laboratories all over the USA since 1970. It is not clear why the EPA changed its opinion on glyphosate from non-cancerous to cancerous in 1991; no other pesticide has seen such a reversal. The EPA's regulations state that the carcinogenicity of a compound should be reviewed whenever sufficient new, relevant information becomes available, but it is difficult to unequivocally identify what kind of new information induced the switch between the 1985 assessment and the 1991 assessment. The EPA's 1991 and (particularly) 2017 standpoints on glyphosate have been heavily criticized, although some of these critical stances are likely a consequence of the ongoing worldwide debate on glyphosate, leading to the obscuring of real facts and the generation of "fake news."

In Europe, the first instance responsible for the evaluation of the glyphosate dossier is the *European Food Safety Authority (EFSA)*. Glyphosate was first registered for the EU market on July 1, 2002, following a review of the 1995 application by Monsanto and others in the Glyphosate Task Force 1. The EU's procedure was followed in all the details: The Draft Assessment Report (DAR) was produced by the evaluating country, Germany, although it is not clear who, in Germany, performed the review. The conclusion of the review was that glyphosate has no carcinogenic properties, and after a review by all member states, glyphosate was approved for the EU market. Such approvals are time-limited, and glyphosate-producing companies requested that the formulation's registration be renewed based on a new dossier. In 2012, Germany was again the reporting country (termed the "rapporteur member state" or RMS). This time, the evaluation was performed by the *Bundesinstitut für Risikobewertung* (BfR), a government scientific organization founded in 2002. The report on glyphosate was released on December 11, 2013 (https://www.bfr.bund.de/cm/349/bfr-contribution-to-the-eu-approval-process-of-glyphosate-is-finalised.pdf). The BfR's report was transmitted to EFSA, which concluded that "glyphosate is unlikely to pose a carcinogenic hazard to humans" and proposed "a new safety measure that will tighten the control of glyphosate residues in food." The Authority organized a meeting in 2015 on the toxic properties of glyphosate with competent experts in the field. The conclusions of this meeting were published on 12 November 2015, recognizing, in accordance with EFSA's previously published opinion, that glyphosate should not be categorized as carcinogenic. EFSA proposed an acceptable operator exposure level of 0.1 mg/kg body weight/day and an acceptable daily intake level of 0.5 mg/kg/day. The conclusion of the EFSA pesticide peer review is published in the EFSA Journal and is available on the EFSA website (https://efsa.onlinelibrary.wiley.com/doi/epdf/10.2903/j.efsa.2015.4302). The EU Regulations related to the glyphosate dossier are depicted in Chap. 11, Annex 2.

The second important European organization to express its point of view on the toxicity of glyphosate was *the European Chemicals Agency (ECHA)*. The ECHA is responsible for the correct classification, labeling, and packaging of chemicals. The conclusion of ECHA is in line with the conclusion of EFSA: There is no evidence for the carcinogenic, mutagenic, or reproductive toxicity of glyphosate. However, the ECHA classified glyphosate as capable of causing eye damage and as severely toxic to aquatic life. The conclusions of the ECHA can be found on the ECHA website (https://echa.europa.eu/-/glyphosate-not-classified-as-a-carcinogen-by-echa).

To comment and conclude on the EFSA and ECHA, similar to the EPA, the EFSA and ECHA have gained a solid scientific reputation, although concern has been raised over the independence of some EFSA employees. However, the methodology followed by EFSA in vetting its employees for ties with other, possibly compromising, organizations is very strict and publicly available. The EFSA and the expert panel identified a number of gaps in the available literature, but, according to EFSA, these gaps are related mainly to laboratory methods and do not prevent EFSA from making their conclusion.

The methodology used by EFSA has been described in a 152-page document entitled "Programming document 2024–2026; science, safe food and sustainability" (https://www.efsa.europa.eu/sites/default/files/2024-01/programming-document-2024-2026.pdf). The document mainly illustrates the workload of EFSA due to the growing sophistication of scientific methods and results. Related to the actual permit to market glyphosate till 2033, one would expect that, given the contested conclusions on the further use of glyphosate and given the amount of new information that became available between 2020 and 2023, a cautious attitude vis-à-vis glyphosate would have been indicated and appreciated. However, "glyphosate" is mentioned only once (p. 49 in this document); the notion of "precautionary principle" is not mentioned. This leaves little room for discussion and strenghten a feeling of disappointment.

The fourth organization is the *International Agency for Research on Cancer (IARC)*. The IARC was founded on 20 May 1965 as the specialized cancer agency of the WHO, with its headquarters in Lyon. In 2015, 17 independent scientists published monograph 112, entitled "Carcinogenicity of tetrachlorvinphos, parathion, malathion, diazinon, and glyphosate." Glyphosate is considered to be a Class 2A carcinogen, meaning that it is "probably carcinogenic to humans" (compounds classified as 2B are "possibly carcinogenic to humans"). IARC's explanation is as follows: "*There is limited evidence of carcinogenicity in humans and sufficient evidence in experimental animals. Exceptionally, an agent (or mixture) may solely be classified under this category if there is limited evidence of carcinogenicity in humans, but if it clearly belongs to this category on the basis of mechanistic considerations.*" Furthermore, IARC explains that "limited evidence" is assigned if "*a positive association has been observed between exposure to the agent and cancer for which a causal interpretation is considered by the Working Group to be credible, but chance, bias or confounding could not be ruled out with reasonable confidence.*" These conclusions were published first as a short note in Lancet Oncology (https://www.iarc.fr/wp-content/uploads/2018/07/MonographVolume112-1.pdf) and in full on 20/3/2015 (https://monographs.iarc.fr/wp-content/uploads/2018/07/mono112.pdf). The animal carcinogenicity analysis of IARC was published in Extenso later (Portier, 2020).

The IARC manual describes its methodology as follows: "*The IARC Monographs Program seeks to classify cancer hazards, meaning the potential of any substance to cause cancer based on current knowledge. The classification does not indicate the level of risk that exists to people's health associated with exposure to a classified hazard. For example, the IARC has classified tobacco smoking as carcinogenic to humans (Group 1), but that classification does not indicate an increase in risk for each cigarette smoked.*"

The IARC reviewed nearly 1000 compounds in 125 monographs. Most of them are classified as Classes 1, 2A, or 2B, meaning they are carcinogenic or should be treated as such. Only one compound in the list (caprolactam) was noncarcinogenic (Class 4). The full list of compounds investigated by the IARC can be found at https://monographs.iarc.fr/list-of-classifications-volumes/.

The IARC uses published peer-reviewed information only; unpublished material is not considered. The mission statement of the IARC mentions that "*the Monographs do not necessarily cite all the literature concerning the subject of an evaluation. Only those data considered by the Working Group to be relevant to making the evaluation are included. With respect to biological and epidemiological data, only reports that have been published or accepted for publication in the openly available scientific literature are reviewed by the working groups. In certain instances, government agency reports that have undergone peer review and are widely available are considered. Exceptions may be made on an ad hoc basis to include unpublished reports that are in their final form and publicly available if their inclusion is considered pertinent to making a final evaluation. In the sections on chemical and physical properties and on production, use, occurrence and analysis, unpublished sources of information may be used."*

The aim of the IARC is to identify carcinogens on a biomechanistic basis, focusing on the hazard of the compound and less on the risk for human toxicity. Whether the pure compound or a formulation is used is a point of discussion. Animal studies in vivo use the pure compound or an accepted formulation to perform the tests; in the case of glyphosate, MON52276; this point is discussed in detail in Chap. 9. Notably, human toxicity data are always obtained using formulations of glyphosate, such as Roundup, which contains adjuvants. IARC concludes that the extent of exposure to glyphosate in real-field circumstances, i.e., when formulations containing glyphosate are used, is of the same magnitude as what is used in experimental conditions with pure glyphosate. Studies of humans exposed to different formulations in different regions at different times are suggestive of a similar increase in the same type of cancer, non-Hodgkin lymphoma. Data on pure glyphosate from animal and other experimental studies, including studies on human cells, support the conclusions of studies on exposed people. For studies of pure glyphosate, the Monograph revealed that the evidence for cancer in experimental animals was "sufficient" and that the evidence for genotoxicity was "strong."

IARC is correct in its assessment that glyphosate is carcinogenic; according to the results of biological in vitro experiments and in vivo animal experiments, the evidence for human carcinogenicity is less convincing. The conclusions of the IARC Monograph have been used as a reason to restrict the use or ban glyphosate formulations in some countries.

To comment and conclude on the IARC, the IARC has a global influence in discussions in cancer research. The scientific community agrees with IARC's carcinogenic classification for the vast majority of the examined compounds, but for some compounds, this is less clear. Monograph 51 (from 1991) classifies coffee as a 2A compound ("probably carcinogenic to humans"), reclassifying it in 2018 to 2B in Monograph 116. Several natural products frequently present in healthy diets have been classified as 2A or 2B. Furthermore, the mission statement on the use of information (see above) is not clear about the criteria for choosing the information, which leaves room for selection bias.

Taken together, worldwide renowned scientific organizations conclude differently about the carcinogenicity of glyphosate, and the key question is why. There are several reasons for this.

First, they do not use the same data (Benbrook, 2019; Tarazona et al., 2017). The IARC uses only peer-reviewed, published data; the EPA and EFSA also use unpublished data, provided they are of high scientific quality. The debate on what data is admissible or not was very negatively affected by stakeholders and the media accusing the organizations of omitting valuable information. These accusations are usually not published in the regular scientific circuit, with some exceptions. Recall the statement of the IARC (see above): "Only those data considered by the Working Group to be relevant to making the evaluation are included"; this could lead to selection bias. Comparing the conclusions of the three organizations is thus similar to comparing apples and oranges.

In addition, when new relevant information becomes available, it should be possible to formulate a new point of view. However, this does not happen—at least not frequently (see the example of coffee), and new information is sometimes disregarded. One of the most influential papers on this topic was from Seralini et al. (2012), who described a significant increase in liver and kidney necrosis and mammary tumors in female rats after glyphosate exposure. The paper was heavily disputed and was ultimately retracted from publication. Notably, the retracted paper of Séralini, including the original data, was accepted for publication in 2015 (Mesnage et al., 2015). In 2018, 3 years after IARC published its monograph on glyphosate, the results of a large epidemiological study in the USA (the Agricultural Health Study—AHS) were published, demonstrating the absence of carcinogenicity in real-world situations (Andreotti et al., 2018). This study was an update of an earlier prospective cohort study that included data from 2010 and 2012. This study concluded that *"In this large, prospective cohort study, no association was apparent between glyphosate and any solid tumours or lymphoid malignancies overall, including non-Hodgkin lymphoma (NHL) and its subtypes*." There was limited evidence of an increased risk of acute myeloid leukemia (AML) among the highest exposure group. Zhang et al. (2019) (odds ratio (OR): 1.41; 95% confidence interval (CI): 1.13–1.75). Hence, whether glyphosate is carcinogenic depends on the set of data used for the analysis. A detailed analysis of the IARC assessment by four independent expert panels concluded that the IARC conclusion is probably wrong (Williams et al., 2016), whereas other papers support the IARC position (Samet, 2015).

Second, there are several essential and inevitable problems with the data. Statements regarding the carcinogenicity of glyphosate to humans under real-life conditions rely on systems for recording cancer and epidemiological studies, which are at least partly qualitative. Although this is the only way to obtain information, epidemiological analysis has intrinsic drawbacks (see Chap. 2). If the collection of data relies on the self-reporting of sick people or of proxies (such as relatives of people who died of cancer), as in case-control retrospective studies, bias is possible. The IARC results have been fundamentally criticized because of recall bias (Crump, 2020; Crump et al., 2020).

Third, economic interests have interfered importantly with the evaluation of glyphosate (McHenry, 2018), and the only way to counteract these malpractices is to apply full transparency throughout the entire evaluation process. This notion has generated some remarkable reactions. Lorenzo Tomatis, the IARC director from 1982 to 1993, accused the IARC of obscuring the risks of industrial chemicals and of "highly irregular" voting procedures, alleging industrial interference (Baines, 2003). In 2003, 30 public health scientists signed a letter targeting conflicts of interest at the IARC and the lack of transparency. The IARC director, Christopher Wild, responded in 2018 to critics with an open letter https://www.iarc.fr/wpontent/uploads/2018/07/IARC_response_to_criticisms_of_the_Monographs_and_the_glyphosate_evaluation.pdf. Infante et al. (2018) demonstrated the interference of economic interests with the IARC and working group members in terms of their position against glyphosate and red meat.

Within this gray zone of conflicts of interest, the issue of confidentiality of economically valuable information interpreted as inherent secrecy of the evaluating organization comes into the picture. In 2018, the EPA administrator, Scott Pruitt, stated, "*The era of secret science at EPA is coming to an end. The ability to test, authenticate, and reproduce scientific findings is vital for the integrity of the rulemaking process. Americans deserve to assess the legitimacy of the science underpinning EPA decisions that may impact their lives.*" This might raise questions about the legitimacy of research conducted at the EPA before 2018. Recall that the position of the EPA on glyphosate dates back to 1985, 1991, 1995, and 2017. In October 2013, the EFSA held a conference with stakeholders to discuss how transparency could be further enhanced in the EFSA's risk assessment process. Mutual accusations of malpractice, conflicts of interest (five papers in Critical Reviews in Toxicology, 2016), industry financial support from researchers to purchase a point of view, evidence published in respected journals but later identified as fraud, the infamous Monsanto papers (McHenry, 2018), and so on are numerous, and all four organizations defended themselves against these accusations with some degree of anger (Brusick et al., 2016, 2018). The funding of the Ramazzini Institute initiative to conduct a 3-year animal study is, according to the organization, based on crowdfunding. It is not clear what exactly this involves or how the results of this study would contribute to the debate, but it is at least a serious statement to remain independent of funding from companies or other organizations.

Finally, the aim of the organizations involved is different. IARC Monograph 51 states, "*The Monographs represent the first step in carcinogenic risk assessment, which involves examination of all relevant information in order to assess the strength of the available evidence that certain exposures could alter the incidence of cancer in humans. The second step is quantitative risk estimation, which is not usually attempted in monographs. Detailed, quantitative evaluations of epidemiological data may be made in the Monographs but without extrapolation beyond the range of the data available. Quantitative extrapolation from experimental data to the human situation is not undertaken.*" EPA and EFSA/ECHA consider exposure to glyphosate in their evaluation.

The conclusion from previous paragraphs is that the situation is, euphemistically speaking, complex. The next question is if the scientific literature could solve (some of) the problems. What follows is a short overview of older and recent peer-reviewed papers.

The number of papers on the link between exposure to pesticides, including glyphosate, is extremely high, and a complete overview is nearly impossible. In general, the animal experiments (2-year rat or mouse daily exposure) yielded clear data according to Portier: Glyphosate is carcinogenic to rodents (Portier, 2020). The human epidemiological data are less clear and yield a mixed picture, although the suggestion of a link between exposure (long exposure or exposure to high doses) and non-Hodgkin lymphoma is suggested. Hence, some doubts about the carcinogenicity of glyphosate remain (Williams et al., 2016; Benbrook, 2019). In what follows are some papers published in the period 2013–2019, and they do not succeed in removing this distrust.

- The open letter of Portier to Vytenis Andriukaitis, EU Commissioner for Agriculture and Human Development, entitled *Review of the Carcinogenicity of Glyphosate by EFSA and BfR,* was signed by 94 scientists.
- http://www.efsa.europa.eu/sites/default/files/Prof_Portier_letter.pdf.
- Portier is a prominent member of the IARC group that works on the glyphosate dossier and Monograph 112, and the authors of this open letter claim to represent "a group of independent and governmental scientists from around the world who have dedicated our professional lives to understanding the role of environmental hazards on cancer risks and human health." The IARC opinion of the carcinogenicity of glyphosate is maintained: Glyphosate belongs to Class 2A.
- Chang and Delzell (2016). *Systematic review and meta-analysis of glyphosate exposure and risk of lymphohematopoietic cancers.*
- In this extensive review, the authors use the available epidemiological studies and conclude that "A causal relationship has not been established between glyphosate exposure and risk of any type of lymphohaematopoietic cancer." The cancers included non-Hodgkin lymphoma, Hodgkin lymphoma, multiple myeloma, and leukemia. Odds ratios were marginally statistically significant for all types of cancers.
- Schinasi and Leon (2014). *Non-Hodgkin lymphoma and occupational exposure to agricultural pesticide chemical groups and active ingredients: a systematic review and meta-analysis.*
- This extensive review includes data from developed countries on 21 groups of pesticides and 80 active compounds. A positive correlation between glyphosate exposure and the occurrence of the B-cell subtype of non-Hodgkin lymphoma was reported.
- Samet (2015) *The IARC monographs: critics and controversy.*
- Samet concludes that "The IARC processes are robust and transparent and not flawed and biased as suggested by some critics." Moreover, the author identified several problems, such as the limited reliance on epidemiological data and the intrinsic limitations of the IARC process. Additionally, the independence of the participants is questioned. The message from this Samet paper is not clear, or at least disputable.
- Davoren and Schiestl (2018). *Glyphosate-based herbicides and cancer risk: a post-IARC decision review of potential mechanisms, policy, and avenues of research.*
- The authors conclude that traditional toxicological tests may no longer be relevant for evaluating the carcinogenicity of substances with complex carcinogenic profiles such as glyphosate. The authors stress that the mode of action of carcinogenicity probably includes microbiome disruption and endocrine mimicry at very low concentrations. The authors conclude that, in view of the evolution of scientific knowledge, particularly of the microbiome, relying on historical data is inadequate.

- Bus (2017). *IARC use of oxidative stress as a key mode of action characteristic for facilitating cancer classification: glyphosate case example illustrating a lack of robustness in interpretative implementation.*
- The author reminds us that the working principles of the IARC include several mechanistic and biomechanistic characteristics or "modes of action" for evaluating the hazard of human carcinogenicity, including the oxidative stress model. According to the authors, the use of oxidative stress as a key mode of action characteristic for facilitating cancer classification is challenged. The authors question the risk–hazard transition of glyphosate performed by the IARC.
- Tarazona et al. (2017). *Glyphosate toxicity and carcinogenicity: a review of the scientific basis of the European Union assessment and its differences with IARC.*
- The authors claim that the actual exposure levels to glyphosate are below toxicological reference values and do not represent a public concern. The differences in the carcinogenicity assessment between the EU and IARC are due to the use of different data sets, particularly on long-term toxicity and carcinogenicity in rodents. This could partially explain the divergent findings, but methodological differences in the evaluation of the available evidence have also been identified (see Table 1 in Tarazona et al., 2017). In addition, the authors highlight the need to develop mechanistic assessments to unravel a possible mode of action. An extensive overview of the available animal studies is presented in support of the conclusion: "*… that actual exposure levels (of glyphosate) are below (toxicological) reference values and do not represent a public concern.*" Notably, the first author of this paper is an employee of EFSA.
- Clausing et al. (2018). *Pesticides and public health: an analysis of the regulatory approach to assessing the carcinogenicity of glyphosate in the European Union.*
- The authors question the classification of glyphosate by the EFSA because the agency did not follow its own application or guideline documents for evaluating carcinogenicity hazards. In particular, the authors conclude that the statistical analysis (one-tailed versus two-tailed tests in carcinogenicity issues) disregards the principle of dose–response relationships, which jeopardizes the conclusions. They conclude that the IARC's classification is justified.
- Landrigan and Belpoggi (2018). *Independent research on the health effects of glyphosate-based herbicides is needed.*
- The authors are members of the Ramazzini Institute, an independent cancer research institute. The organization is setting up a three-year prospective study on the animal carcinogenicity of glyphosate.
- Manservisi et al. (2019). *The Ramazzini Institute 13-week pilot study glyphosate-based herbicides administered at human-equivalent dose to Sprague Dawley rats: effects on development and endocrine system.*
- This pilot is the predecessor of the planned prospective three-year study mentioned above, which emphasizes the endocrine deregulation caused by glyphosate. Although a clear effect on reproductive development parameters in male and female rats was observed at one dose, these findings need confirmation.

- Mao et al. (2018). *The Ramazzini Institute 13-week pilot study on glyphosate and roundup administration at human-equivalent dose to Sprague Dawley rats: effects on the microbiome.*
- This is a parallel study to Manservisi's 2019 study focusing on the microbiome. The effect of glyphosate on the rat gut microbiota at an early developmental stage, particularly before the onset of puberty, was evident.
- Torretta et al. (2018). *This article provides a critical review of the effects of glyphosate exposure on the environment and humans through the food supply chain.*
- The authors conclude: "*The results have brought to light how massive use of the herbicide has created over time real global contamination that has not only affected the soil, surface and groundwater as well as the atmosphere but even food and commonly used objects such as diapers, medical gauze and others. Despite numerous studies regarding the dangers resulting from the extensive use of glyphosate, it is not possible to attribute an unambiguous classification to glyphosate, especially regarding its potentially harmful effects on humans.*"
- Ward, (2018). *Glyphosate use and cancer incidence in the agriculture health study: an epidemiological perspective.*

- In this updated analysis of a long-term agriculture health study in the USA (the "AHS" study), an increased but statistically insignificant risk of acute myelogenous leukemia was found in the quartile with the highest exposure (RR = 2.44; CI = 0.94–6.32). The authors noted that the AHS is one of the largest epidemiological studies ever on glyphosate, including 57,310 licensed users and 32,347 spouses.

Some more recent studies show the same tendency.

- Chang et al. (2023). The authors conclude that "… the weight of evidence supports an association between glyphosate exposure and oxidative stress in humans." The evidence that oxidative stress has carcinogenic potential is well known.
- Rezende et al. (2021). The authors conclude that glyphosate below the official safety limits has toxic effects, but the scientific data are not concise enough for disapproval.
- Franke et al. (2021). In a study with postmenopausal women with 124 breast cancer patients and 126 healthy controls, there was no link between urinary glyphosate levels and the occurrence of tumors.
- Crump et al. (2020). The authors consider all glyphosate-related tumors as false positives because of the incorrect application of statistical methods by the IARC. They concluded that during the evaluation process, multiple comparisons are made, making statistical significance more probable but meaningless. Notably, Crump was involved in the EPA Federal Insecticide, Fungicide and Rodenticide Act and in the glyphosate review in 2016. Haseman had an agreement with Monsanto in 2017 (https://hh-ra.org/wp-content/uploads/Crump-

Accounting-for-Multiple-Comparisons-in-S.pdf). In this case, some reservations are at stake.

The purpose of showing this list of papers is merely to highlight contradictions, not completeness. It looks as if for every paper denying carcinogenicity, you can find another paper accepting carcinogenicity, with all papers claiming to be correct.

Even if you accept that glyphosate is carcinogenic, it seems that its carcinogenicity is not comparable to the carcinogenicity of known carcinogens such as smoking or asbestos; it is a weak carcinogen. However, being a "weak or strong" carcinogen is not easy to define. Whether a compound is a weak or strong carcinogen is based on in vivo studies in animals and epidemiological studies in humans and not on biological mechanisms (Schrenk, 2018). This implies that the exposure level and body burden of the carcinogen are crucial in the discussion. Hence, if so, it might be possible to establish the lowest observed adverse effect level (LOAEL) for tumorigenic endpoints and derive admissible exposure levels.

As mentioned in Chap. 2, this is not possible for genotoxic compounds for which no safe exposure level is possible. The EU approached the issue in 1999, setting exposure limits to distinguish between weak and strong carcinogens (http://ec.europa.eu/environment/archives/dansub/pdfs/potency.pdf). DNA damage and methylation in human peripheral mononuclear cells have been reported (Kwiatkowska et al., 2017), although the possibility that the adjuvant causes genotoxicity has been proposed (Brusick et al., 2018). Under real-life conditions, DNA damage was detected in an Ecuadorian population upon exposure to the commercial preparation of glyphosate containing the adjuvant (Paz-Y-Mino et al., 2007). Here, again, controversial options have been published denying genotoxicity (Wozniak et al., 2018). Nagy et al. reported evidence of genotoxicity in human peripheral blood cells when exposed to both glyphosate and tallowamine (Nagy et al., 2019). The same group repeated the experiments with glyphosate alone and reached the same conclusion: Glyphosate by itself is genotoxic (Nagy et al., 2021). Numerous other papers with similar conclusions have been published (Singh et al., 2024; Wozniak et al., 2018, 2020). These results contradict those of others who reported alterations in sister chromatid exchange and chromosomal aberration in cultured human blood cells only at the highest concentration of glyphosate (200 μM), which, according to the authors, is irrelevant to the real situation on the ground (Abu Tarboush et al., 2022). Here, again, controversial results proper to scientific research have been published. Taken together, the issue of weak and strong carcinogens, together with exposure to multiple carcinogens, and the possible interaction with cellular genetic material, is complex, and the interpretation of epidemiological data should be performed with great care (Yamaguchi, 1999).

How can a neutral observer form his or her own opinion?

One possible route is to follow the Bradford Hill criteria for investigating causality (Hill, 1965). The evaluation of the criteria in relation to glyphosate is as follows:

- Criteria 1 (strength of association between possible cause and possible consequence) and 2 (consistency): The association between the cause (exposure to glyphosate) and the proposed consequence (cancer in humans) is at least disputable. Even the largest epidemiological studies are not unequivocal except at the highest levels of exposure.

- Concerning criterion 3 (specificity) and criterion 6 (plausibility): There is no doubt that we are exposed to mixtures of chemicals, some of which are known carcinogens (such as car exhaust and cigarette smoke). Since glyphosate is present in all compartments of the environment, it is part of this mixture; linking a particular cancer type to glyphosate is therefore difficult. This is also the case for non-Hodgkin lymphomas in general, although there is a slight indication that the B-cell lymphoma subtype might be linked to glyphosate exposure. The importance of identifying subgroups in populations with increased sensitivity to glyphosate is obvious, although this aspect is largely unknown (Manach et al., 2017).
- Criterion 4 (temporality): It is tempting to associate the development of certain types of cancer with exposure to glyphosate, although the assumed length of the latency period hampers any conclusion. There are numerous ongoing lawsuits against Monsanto and Bayer in the USA; in particular, in the case of Dewayne Johnson, a temporal link between the long-term use of glyphosate and cancer was accepted by a federal judge. Notably, for other non-cancerous diseases such as Parkinson's disease and other neuronal diseases, this temporal link is also suggested.
- Criteria 5 (biological gradient), 7 (coherence), and 8 (experiment) refer to the biological basis of the assumed cancer-causing agent. The results of in vitro experiments yield evidence for the carcinogenicity of glyphosate. However, the doses and concentrations needed to induce the cancer phenotype in experiments are, as a rule, many times greater than what is generally observed in in vivo conditions during human exposure, whereas in classical 2-year carcinogenicity studies, the spontaneous appearance of tumors in rodents hampers any conclusion on the assumed glyphosate-induced tumors. These results have some predictive value for the genotoxicity or epigenetic transformation, although extrapolation to the human situation is speculative.
- Criterion 9 does not apply since there is no similar compound to glyphosate for comparison.

Taken together, the Bradford–Hill criteria are fulfilled to some degree, although additional criteria, such as dose–response conditions and low-dose exposures, are not fulfilled. Other effects than cancer, such as effects on the microbiome, occur at very low exposure levels comparable to those seen in humans. The straightforward application of the Hill criteria has been questioned, given advancements in determining the mode of action and (bio)medical background (Gross & Fedak, 2015).

How can the epidemiological data on pesticides, including glyphosate, be collected?

For an introduction to epidemiology, see Chap. 2. Most recently, Gerken et al. (https://www.frontiersin.org/journals/cancer-control-and-society/articles/10.3389/fcacs.2024.1368086/full) used freely available data from governmental organizations such as the United States Geological Survey for pesticide data, the National Institute of Health and the Centres for Disease Control for Cancer Data and agricultural land use data from the US Department of Agriculture. In total, data for 69 pesticides (including glyphosate) were obtained and statistically analyzed via latent class analysis. After correcting for confounding data (smoking and socioeconomic vulnerability), they concluded that exposure to pesticides is linked to leukemia, non-Hodgkin lymphoma, bladder, colon, lung, and pancreatic cancers and all cancers combined with smoking. The authors even suggest that, when a property is bought, the potential buyer must be informed of the use of pesticides in the vicinity of the property. Similarly, Zhou et al. summarized the human health impacts of chemical pesticide use (Zhou et al., 2025) and glyphosate (Galli et al., 2024; Singh et al., 2024). Here, again, the conclusion is straightforward: Pesticides contribute significantly to endangering human health in the context of cancer, neurological problems, etc.

The EPA discussed the data on epidemiological evidence for the carcinogenicity of glyphosate in 2016 (https://www.epa.govt.nz/assets/Uploads/Documents/Everyday-Environment/Publications/EPA-glyphosate-review.pdf). Overall, the odds ratios (see the next paragraph) reflecting the risk of having NHL or any other type of cancer were all close to 1, with 2.1 being the highest value. These findings support the IARC conclusion that the evidence for carcinogenicity in humans is "limited." In an Italian multicenter case-control study, an overall increasing trend for lymphoma subtypes with increasing exposure duration was found (Meloni et al., 2021). Reflecting the available data to the Bradford Hill criteria, Weisenburger concluded that a link between glyphosate exposure and the occurrence of non-Hodgkin lymphoma was evident (Weisenburger, 2021).

Glyphosate Is a Weak Carcinogen

Fundamental questions concerning chemical carcinogenicity and the actual approach were discussed in a series of important papers (Cohen et al., 2019; Doe et al., 2019; Wolf et al., 2019). Many studies have confirmed the increase in biological markers of oxidative stress and inflammation after glyphosate exposure, both in vivo in animals and in vivo in humans (Chang et al., 2023; Makris et al., 2022; Milic et al., 2018; Sapbamrer et al., 2019; Sidthilaw et al., 2022; Sosa et al., 2013). This wealth of experimental data cannot be neglected. The question is how to interpret these data in one coherent, acceptable overview.

The risk of developing cancer upon exposure to a substance is frequently expressed as an odds ratio. This figure compares the possibility (the odds) of having cancer in an exposed group with those in a non-exposed control group. This ratio can be =1, <1, or >1, meaning no, a negative, or a positive association, respectively. Some data are summarized in the next table.

<table>
<tr><th>References</th><th>Agent</th><th>Type of cancer</th><th colspan="3">Odd's ratio's + confidence intervals (95% or interval)</th></tr>
<tr><td rowspan="4">Pesch et al. (2012)</td><td rowspan="4">Smoking (>30 cigarettes/day)</td><td rowspan="4">Lung cancer</td><td></td><td>Male</td><td>Female</td></tr>
<tr><td>Squamous cell carcinoma.</td><td>103.5 (74.8)</td><td>62.7 (31.5)</td></tr>
<tr><td>Small cell lung cancer</td><td>111.3 (69.8)</td><td>108.6 (50.7)</td></tr>
<tr><td>Adenocarcinoma</td><td>21.9 (16.6)</td><td>16.8 (9.2)</td></tr>
<tr><td>Powell et al. (2013)</td><td>Smoking</td><td>Lung cancer</td><td colspan="3">Male: 12.81 (11.52–14.24), female: 19.1 (16.98–21.49)</td></tr>
<tr><td>Villeneuve et al. (2011)</td><td>Smoking</td><td>Lung carcinoma's</td><td colspan="3">40 years of smoking: 21.6 (14.66–31.84)</td></tr>
<tr><td>Frederiksen et al. (2020)</td><td>Parental smoking</td><td>Acute lymphoblastic leukemia acute myeloid leukemia</td><td colspan="3">Smoking before conception: 2.51 (1.21–5.17).
Smoking during pregnancy: 3.21 (1.56–6.6).
Smoking after birth: 2.83 (1.36–5.9)</td></tr>
<tr><td>Villeneuve et al. (2011)</td><td>Diesel exhaust</td><td>Squamous and large cell lung cancer</td><td colspan="3">Cumulative lifetime exposure: 1.34 (1.11–1.60)
Highest exposure concentrations: 1.68 (1.17–2.35)</td></tr>
<tr><td>Koutros et al. (2020)</td><td>Diesel exhaust</td><td>Bladder cancer</td><td colspan="3">1.61 (1.08–2.4)</td></tr>
<tr><td>Talibov et al. (2019)</td><td>Diesel exhaust</td><td>Colorectal cancer</td><td colspan="3">1.05 (1.02–1.08)</td></tr>
<tr><td>Zeng et al. (2019)</td><td>Alcohol</td><td>Hypopharyngeal cancer</td><td colspan="3">11.56 (6.53–19.41)</td></tr>
<tr><td>Diallo et al. (2018)</td><td>Red and processed meat</td><td>Cancer</td><td colspan="3">Breast cancer: 1.83 (1.33–2.51), overall cancers: 1.31 (1.10–1.55)</td></tr>
<tr><td>Ferro et al. (2020)</td><td>Red meat</td><td>Gastric cancer</td><td colspan="3">Total meat (red + processed): 1.30 (1.09–1.55)</td></tr>
<tr><td>Liu et al. (2012)</td><td>Aflatoxin</td><td>Hepatocellular cancer</td><td colspan="3">6.37 (3.74–10.86) (odd's ratio's increase in HBV+ patients to 73.0 (36–148))</td></tr>
<tr><td>Parent et al. (2003)</td><td>Asbestos</td><td>Mesothelioma</td><td colspan="3">15.0 (2.9–78.6)</td></tr>
<tr><td>West et al. (2019)</td><td>Asbestos</td><td>Parenchymal and pleural abnormalities</td><td colspan="3">>50 years work in asbestos milieu: 26.65 (18.46–38.46)</td></tr>
<tr><td>'t Mannetje et al. (2011)</td><td>Metal compounds</td><td>Lung cancer</td><td colspan="3">Arsenic: 1.65 (1.05–2.58), chromium: 1.25 (0.95–1.65) and cadmium 1.19 (0.77–1.82)</td></tr>
<tr><td>Ferreccio et al. (2013)</td><td>Arsenicum</td><td>Bladder cancer</td><td colspan="3">Combination of arsenic exposure and smoking: 16.95 (6.5–40) for lung cancer and 23 (8.2–66) for bladder cancer</td></tr>
<tr><td>Xu et al. (2015)</td><td>Urinary N-nitroso compounds</td><td>Gastric cancer</td><td colspan="3">3.27 (0.76–14.4) and 4.82 (1.05–22.17) for second and third tertile of urinary concentration of nitroso compounds</td></tr>
</table>

(continued)

References	Agent	Type of cancer	Odd's ratio's + confidence intervals (95% or interval)
Andreotti et al. (2018)	Glyphosate	Acute myeloid leukemia	2.44 (0.94–6.32)
Andreotti et al. (2019)	Glyphosate highest exposure group Q3 and Q4	Acute myeloid leukemia	Q3: 2.70 (1.14–6.43) Q4: 2.62 (1.14–6.07)
Eriksson et al. (2008)	Herbicides, glyphosate	Non-Hodgkin lymphoma	Herbicides (not specified): 1.72 (1.18–2.51) Glyphosate: 2.02 (1.10–3.71)
Pahwa et al. (2019)	Glyphosate	Non-Hodgkin lymphoma	All subtypes of non-Hodgkin lymphoma: 1.43 (1.11–1.83) Diffuse large B-cell lymphoma: 2.14 (1.07–4.28)

References to the Table

Lung cancer and smoking (Pesch et al., 2012; Powell et al., 2013; Stapelfeld et al., 2020); parental smoking (Frederiksen et al., 2020); diezel and lung cancer (Koutros et al., 2020; Talibov et al., 2019; Villeneuve et al., 2011); alcohol and hypopharyngeal cancer (Zeng et al., 2019); red meat and gastric cancer (Diallo et al., 2018; Ferro et al., 2020); aflatoxin and hepatocellular cancer (Liu et al., 2012); asbestos (Parent et al., 2003; West et al., 2019); metal compounds ('t Mannetje et al., 2011); arsenic and bladder cancer (Ferreccio et al., 2013); N-nitroso compounds (Xu et al., 2015); pesticides including glyphosate and non-Hodgkin lymphoma (Andreotti et al., 2018, 2019; Erickson et al., 2023; Eriksson et al., 2008; Mesnage et al., 2019; Pahwa et al., 2019).

Numerous other papers confirm these data for smoking (Rahal et al., 2017; Steliga & Dresler, 2011) and for asbestos (Furuya et al., 2018; Odgerel et al., 2017). For other papers, careful inspection is indicated. In the paper of Leon et al. (2019), the ODD ratio for non-Hodgkin lymphoid malignancies was moderately elevated to 1.36. Notably, 5/17 of the authors of the study were employees of the IARC.

This table requires additional comments.

First, although odd ratios indicate a carcinogenic risk, some are heavily disputed. Moderate alcohol consumption is considered harmless and even beneficial by various alcohol industrial associations or other groups with political and economic interests. Admittedly, alcohol-dependent cancers associated with excessive alcohol consumption are frequently caused by interference with unhealthy lifestyle elements obscuring the role of alcohol. Red meat is classified by the IARC in Group 1 (carcinogenic to humans), but other organizations are less convinced, including INSERM and the American Cancer Society; in the latter, moderate consumption is allowed (or even advised). The carcinogenicity of red meat is also contested by the American Meat Institute and other organizations with economic motives. There is less doubt about processed meat, which is generally considered carcinogenic. Taken together, the figures from the table need cautious interpretation.

Second, the above list is not an exhaustive literature search but rather a selection used for comparative purposes. Nevertheless, these data suggest the existence of *three levels of cancer risk* dependent on the compound. The highest risk level or "strong" carcinogen is for smoking and asbestos, and this is beyond any reasonable doubt (whatever lobbying groups might tell); there is a middle level for several compounds, such as aflatoxin, processed meat, and arsenic, and there is a low level or "weak" carcinogen for diesel exhaust, red meat, n-nitroso compounds, and glyphosate. The conclusion is that glyphosate is a WEAK carcinogen compared with "established" carcinogens such as smoking and asbestos.

This raises an inevitable question. If it is true that the risk of developing cancer from glyphosate is comparable to that of other daily products, why ban the pesticide? However, this is a very weak argument. The exposure to many of the compounds mentioned in the table is a personal choice. It is our own choice to eat red or processed meat, and we decide for ourselves whether we smoke or not. Hence, the risk of exposure to a carcinogenic compound is a deliberate choice. In contrast, children living in the vicinity of smokers do not have a choice: they are exposed to smoke and have an increased risk of developing cancer. This is also the case for glyphosate: It is used in such quantities, and it is found in all compartments of the environment that we are unwillingly exposed to glyphosate; there is no choice. As for smoking, it is unethical to expose people unwillingly to glyphosate.

Possible Biological Mechanisms to Explain the Carcinogenicity of Glyphosate

A unique target organ, tissue, or cell biological mechanism within mammalian cells that is responsible for the carcinogenicity of glyphosate is difficult to pinpoint; see also the previous paragraph on genotoxicity. Other non-genetic targets have been suggested. Cavalli et al. (Cavalli et al., 2013) demonstrated that the effect of low-dose glyphosate on pre-puberal Sertoli cells in rats was due to calcium overload, which in turn was due to an effect of glyphosate on protein kinase C. This enzyme is a key enzyme in many cellular processes, including signal transduction and growth. Most of this knowledge is based on ex vivo experimental work with isolated cells or systems. The extrapolation to the human whole-body situation therefore remains speculative; see the chapter on risk, hazard, and exposure.

The overall conclusion of this chapter is that glyphosate is a carcinogen, as demonstrated in the laboratory, and is a weak carcinogen for humans in the daily practical use of the herbicide. We should not compare glyphosate to other known carcinogens, such as smoking or asbestos, or to eating processed or red meat. The latter are personal choices, and exposure to glyphosate is not a choice. We cannot ignore the effect of glyphosate because of its omnipresence in all compartments of the environment. The carcinogenicity of glyphosate and, in general, the carcinogenicity of pesticides or chemicals is a real threat to many humans. A precautionary approach is indicated. A multidisciplinary approach including agriculture, chemistry, and biomedical sciences, as well as ethics and sociology, is needed (Kripke et al., 2020). In particular, this research must be performed independently of any political or economic influence (Kogevinas, 2019).

References

't Mannetje, A., Bencko, V., Brennan, P., Zaridze, D., Szeszenia-Dabrowska, N., Rudnai, P., et al. (2011). Occupational exposure to metal compounds and lung cancer. Results from a multicenter case-control study in Central/Eastern Europe and UK. *Cancer Causes and Control, 22*(12), 1669–1680. https://doi.org/10.1007/s10552-011-9843-3

Abu Tarboush, N., Almomani, D. H., Khabour, O. F., & Azzam, M. I. (2022). Genotoxicity of glyphosate on cultured human lymphocytes. *International Journal of Toxicology, 41*(2), 126–131. https://doi.org/10.1177/10915818211073514

Andreotti, G., Koutros, S., Hofmann, J. N., Sandler, D. P., Lubin, J. H., Lynch, C. F., et al. (2018). Glyphosate use and cancer incidence in the agricultural health study. *Jnci-Journal of the National Cancer Institute, 110*(5), 509–516. https://doi.org/10.1093/jnci/djx233

Andreotti, G., Lubin, J. H., Koutros, S., Hofmann, J. N., Sandler, D. P., Lerro, C. C., et al. (2019). Re: Glyphosate use and cancer incidence in the agricultural health study response. *Jnci-Journal of the National Cancer Institute, 111*(2), 216–218. https://doi.org/10.1093/jnci/djy201

Baines, C. J. (2003). Transparency at the International Agency for Research on Cancer (IARC). *Lancet, 361*(9359), 781–782. https://doi.org/10.1016/s0140-6736(03)12635-9

Benbrook, C. (2019). How did the US EPA and IARC reach diametrically opposed conclusions on the genotoxicity of glyphosate-based herbicides? *Environmental Sciences Europe, 31*, 16. https://doi.org/10.1186/s12302-018-0184-7

Brusick, D., Aardema, M., Kier, L., Kirkland, D., & Williams, G. (2016). Genotoxicity Expert Panel review: Weight of evidence evaluation of the genotoxicity of glyphosate, glyphosate-based formulations, and aminomethylphosphonic acid (Publication with Expression of Concern. See vol. 48, pg. 916, 2018). *Critical Reviews in Toxicology, 46*, 56–74. https://doi.org/10.1080/10408444.2016.1214680

Brusick, D., Aardema, M., Kier, L. D., Kirkland, D. J., & Williams, G. (2018). Genotoxicity Expert Panel review: Weight of evidence evaluation of the genotoxicity of glyphosate, glyphosate-based formulations, and aminomethylphosphonic acid (Expression of Concern of Vol 46, Pg 56, 2016). *Critical Reviews in Toxicology, 48*(10), 916–916. https://doi.org/10.1080/10408444.2018.1522133

Bus, J. S. (2017). IARC use of oxidative stress as key mode of action characteristic for facilitating cancer classification: Glyphosate case example illustrating a lack of robustness in interpretative implementation. *Regulatory Toxicology and Pharmacology, 86*, 157–166. https://doi.org/10.1016/j.yrtph.2017.03.004

Cavalli, V., Cattani, D., Rieg, C. E. H., Pierozan, P., Zanatta, L., Parisotto, E. B., et al. (2013). Roundup disrupts male reproductive functions by triggering calcium-mediated cell death in rat testis and Sertoli cells. *Free Radical Biology and Medicine, 65*, 335–346. https://doi.org/10.1016/j.freeradbiomed.2013.06.043

Chang, E. T., & Delzell, E. (2016). Systematic review and meta-analysis of glyphosate exposure and risk of lymphohematopoietic cancers. *Journal of Environmental Science and Health Part B-Pesticides Food Contaminants and Agricultural Wastes, 51*(6), 402–428. https://doi.org/10.1080/03601234.2016.1142748

Chang, V. C., Andreotti, G., Ospina, M., Parks, C. G., Liu, D. P., Shearer, J. J., et al. (2023). Glyphosate exposure and urinary oxidative stress biomarkers in the Agricultural Health Study. *Jnci-Journal of the National Cancer Institute, 115*(4), 394–404. https://doi.org/10.1093/jnci/djac242

Clausing, P., Robinson, C., & Burtscher-Schaden, H. (2018). Pesticides and public health: An analysis of the regulatory approach to assessing the carcinogenicity of glyphosate in the European Union. *Journal of Epidemiology and Community Health, 72*(8), 668–672. https://doi.org/10.1136/jech-2017-209776

Cohen, S. M., Boobis, A. R., Dellarco, V. L., Doe, J. E., Fenner-Crisp, P. A., Moretto, A., et al. (2019). Chemical carcinogenicity revisited 3: Risk assessment of carcinogenic potential based on the current state of knowledge of carcinogenesis in humans. *Regulatory Toxicology and Pharmacology, 103*, 100–105. https://doi.org/10.1016/j.yrtph.2019.01.017

Crump, K. (2020). The potential effects of recall bias and selection bias on the epidemiological evidence for the carcinogenicity of glyphosate. *Risk Analysis, 40*(4), 696–704. https://doi.org/10.1111/risa.13440

Crump, K., Crouch, E., Zelterman, D., Crump, C., & Haseman, J. (2020). Accounting for multiple comparisons in statistical analysis of the extensive bioassay data on glyphosate. *Toxicological Sciences, 175*(2), 156–167. https://doi.org/10.1093/toxsci/kfaa039

Davoren, M. J., & Schiestl, R. H. (2018). Glyphosate-based herbicides and cancer risk: A post-IARC decision review of potential mechanisms, policy and avenues of research. *Carcinogenesis, 39*(10), 1207–1215. https://doi.org/10.1093/carcin/bgy105

Diallo, A., Deschasaux, M., Latino-Martel, P., Hercberg, S., Galan, P., Fassier, P., et al. (2018). Red and processed meat intake and cancer risk: Results from the prospective NutriNet-Sante cohort study. *International Journal of Cancer, 142*(2), 230–237. https://doi.org/10.1002/ijc.31046

Doe, J. E., Boobis, A. R., Dellarco, V., Fenner-Crisp, P. A., Moretto, A., Pastoor, T. P., et al. (2019). Chemical carcinogenicity revisited 2: Current knowledge of carcinogenesis shows that categorization as a carcinogen or non-carcinogen is not scientifically credible. *Regulatory Toxicology and Pharmacology, 103*, 124–129. https://doi.org/10.1016/j.yrtph.2019.01.024

Erickson, P. A., Chang, V. C., Dagnall, C. L., Teshome, K., Machiela, M. J., Barry, K. H., et al. (2023). Occupational pesticide use and relative leukocyte telomere length in the biomarkers of exposure and effect in agriculture study. *Cancer Research, 83*(7), 2. https://doi.org/10.1158/1538-7445.Am2023-4220

Eriksson, M., Hardell, L., Carlberg, M., & Åkerman, M. (2008). Pesticide exposure as risk factor for non-Hodgkin lymphoma including histopathological subgroup analysis. *International Journal of Cancer, 123*(7), 1657–1663. https://doi.org/10.1002/ijc.23589

Ferreccio, C., Yuan, Y., Calle, J., Benítez, H., Parra, R. L., Acevedo, J., et al. (2013). Arsenic, tobacco smoke, and occupation associations of multiple agents with lung and bladder cancer. *Epidemiology, 24*(6), 898–905. https://doi.org/10.1097/EDE.0b013e31829e3e03

Ferro, A., Rosato, V., Rota, M., Costa, A. R., Morais, S., Pelucchi, C., et al. (2020). Meat intake and risk of gastric cancer in the Stomach cancer Pooling (StoP) project. *International Journal of Cancer, 147*(1), 45–55. https://doi.org/10.1002/ijc.32707

Franke, A. A., Li, X. N., Shvetsov, Y. B., & Lai, J. F. (2021). Pilot study on the urinary excretion of the glyphosate metabolite aminomethylphosphonic acid and breast cancer risk: The Multiethnic Cohort study. *Environmental Pollution, 277*, 7. https://doi.org/10.1016/j.envpol.2021.116848

Frederiksen, L. E., Erdmann, F., Wesseling, C., Winther, J. F., & Mora, A. M. (2020). Parental tobacco smoking and risk of childhood leukemia in Costa Rica: A population-based case-control study. *Environmental Research, 180*. https://doi.org/10.1016/j.envres.2019.108827

Furuya, S., Chimed-Ochir, O., Takahashi, K., David, A., & Takala, J. (2018). Global asbestos disaster. *International Journal of Environmental Research and Public Health, 15*(5), 11. https://doi.org/10.3390/ijerph15051000

Galli, F. S., Mollari, M., Tassinari, V., Alimonti, C., Ubaldi, A., Cuva, C., & Marcoccia, D. (2024). Overview of human health effects related to glyphosate exposure. *Frontiers in Toxicology, 6*, 14. https://doi.org/10.3389/ftox.2024.1474792

Gross, S. A., & Fedak, K. M. (2015). Applying a weight-of-evidence approach to evaluate relevance of molecular landscapes in the exposure-disease paradigm. *BioMed Research International, 2015*, 11. https://doi.org/10.1155/2015/515798

Hill, A. B. (1965). Environment and disease – association or causation. *Proceedings of the Royal Society of Medicine-London, 58*(5), 295. https://doi.org/10.1177/003591576505800503

Infante, P. F., Melnick, R., Vainio, H., & Huff, J. (2018). Commentary: IARC Monographs Program and public health under siege by corporate interests. *American Journal of Industrial Medicine, 61*(4), 277–281. https://doi.org/10.1002/ajim.22811

Kogevinas, M. (2019). Probable carcinogenicity of glyphosate. *Bmj-British Medical Journal, 365*, 2. https://doi.org/10.1136/bmj.l1613

Koutros, S., Kogevinas, M., Friesen, M. C., Stewart, P. A., Baris, D., Karagas, M. R., et al. (2020). Diesel exhaust and bladder cancer risk by pathologic stage and grade subtypes. *Environment International, 135*, 10. https://doi.org/10.1016/j.envint.2019.105346

Kripke, M., Brody, J. G., Hawk, E., Hernandez, A. B., Hoppin, P. J., Jacobs, M. M., et al. (2020). Rethinking environmental carcinogenesis. *Cancer Epidemiology Biomarkers & Prevention, 29*(10), 1870–1875. https://doi.org/10.1158/1055-9965.Epi-20-0541

Kwiatkowska, M., Reszka, E., Wozniak, K., Jablonska, E., Michalowicz, J., & Bukowska, B. (2017). DNA damage and methylation induced by glyphosate in human peripheral blood mononuclear cells (in vitro study). *Food and Chemical Toxicology, 105*, 93–98. https://doi.org/10.1016/j.fct.2017.03.051

Landrigan, P. J., & Belpoggi, F. (2018). The need for independent research on the health effects of glyphosate-based herbicides. *Environmental Health, 17*, 4. https://doi.org/10.1186/s12940-018-0392-z

Leon, M. E., Schinasi, L. H., Lebailly, P., Freeman, L. E. B., Nordby, K. C., Ferro, G., et al. (2019). Pesticide use and risk of non-Hodgkin lymphoid malignancies in agricultural cohorts from France, Norway and the USA: A pooled analysis from the AGRICOH Consortium. *International Journal of Epidemiology, 48*(5), 1519–1535. https://doi.org/10.1093/ije/dyz017

Liu, Y., Chang, C. C. H., Marsh, G. M., & Wu, F. (2012). Population attributable risk of aflatoxin-related liver cancer: Systematic review and meta-analysis. *European Journal of Cancer, 48*(14), 2125–2136. https://doi.org/10.1016/j.ejca.2012.02.009

Makris, K. C., Efthymiou, N., Konstantinou, C., Anastasi, E., Schoeters, G., Kolossa-Gehring, M., & Katsonouri, A. (2022). Oxidative stress of glyphosate, AMPA and metabolites of pyrethroids and chlorpyrifos pesticides among primary school children in Cyprus. *Environmental Research, 212*, 8. https://doi.org/10.1016/j.envres.2022.113316

Manach, C., Milenkovic, D., Van de Wiele, T., Rodriguez-Mateos, A., de Roos, B., Garcia-Conesa, M. T., et al. (2017). Addressing the inter-individual variation in response to consumption of plant food bioactives: Towards a better understanding of their role in healthy aging and cardiometabolic risk reduction. *Molecular Nutrition & Food Research, 61*(6). https://doi.org/10.1002/mnfr.201600557

Manservisi, F., Lesseur, C., Panzacchi, S., Mandrioli, D., Falcioni, L., Bua, L., et al. (2019). The Ramazzini Institute 13-week pilot study glyphosate-based herbicides administered at human-equivalent dose to Sprague Dawley rats: Effects on development and endocrine system. *Environmental Health, 18*. https://doi.org/10.1186/s12940-019-0453-y

Mao, Q. X., Manservisi, F., Panzacchi, S., Mandrioli, D., Menghetti, I., Vornoli, A., et al. (2018). The Ramazzini Institute 13-week pilot study on glyphosate and Roundup administered at human-equivalent dose to Sprague Dawley rats: Effects on the microbiome. *Environmental Health, 17*. https://doi.org/10.1186/s12940-018-0394-x

McHenry, L. B. (2018). The Monsanto Papers: Poisoning the scientific well. *International Journal of Risk & Safety in Medicine, 29*(3–4), 193–205. https://doi.org/10.3233/jrs-180028

Meloni, F., Satta, G., Padoan, M., Montagna, A., Pilia, I., Argiolas, A., et al. (2021). Occupational exposure to glyphosate and risk of lymphoma: Results of an Italian multicenter case-control study. *Environmental Health, 20*(1), 8. https://doi.org/10.1186/s12940-021-00729-8

Mesnage, R., Defarge, N., de Vendômois, J. S., & Séralini, G. E. (2015). Potential toxic effects of glyphosate and its commercial formulations below regulatory limits. *Food and Chemical Toxicology, 84*, 133–153. https://doi.org/10.1016/j.fct.2015.08.012

Mesnage, R., Benbrook, C., & Antoniou, M. N. (2019). Insight into the confusion over surfactant co-formulants in glyphosate-based herbicides. *Food and Chemical Toxicology, 128*, 137–145. https://doi.org/10.1016/j.fct.2019.03.053

Milic, M., Zunec, S., Micek, V., Kasuba, V., Mikolic, A., Lovakovic, B. T., et al. (2018). Oxidative stress, cholinesterase activity, and DNA damage in the liver, whole blood, and plasma of Wistar rats following a 28-day exposure to glyphosate. *Arhiv Za Higijenu Rada I Toksikologiju-Archives of Industrial Hygiene and Toxicology, 69*(2), 154–168. https://doi.org/10.2478/aiht-2018-69-3114

Nagy, K., Tessema, R. A., Budnik, L. T., & Adám, B. (2019). Comparative cyto- and genotoxicity assessment of glyphosate and glyphosate-based herbicides in human peripheral white blood cells. *Environmental Research, 179*, 7. https://doi.org/10.1016/j.envres.2019.108851

Nagy, K., Tessema, R. A., Szász, I., Smeirat, T., Al Rajo, A., & Adám, B. (2021). Micronucleus formation induced by glyphosate and glyphosate-based herbicides in human peripheral white blood cells. *Frontiers in Public Health, 9*, 12. https://doi.org/10.3389/fpubh.2021.639143

Odgerel, C. O., Takahashi, K., Sorahan, T., Driscoll, T., Fitzmaurice, C., Yoko-o, M., et al. (2017). Estimation of the global burden of mesothelioma deaths from incomplete national mortality data. *Occupational and Environmental Medicine, 74*(12), 851–858. https://doi.org/10.1136/oemed-2017-104298

Pahwa, M., Freeman, L. E. B., Spinelli, J. J., Blair, A., McLaughlin, J. R., Zahm, S. H., et al. (2019). Glyphosate use and associations with non-Hodgkin lymphoma major histological sub-types: Findings from the North American Pooled Project. *Scandinavian Journal of Work Environment & Health, 45*(6), 600–609. https://doi.org/10.5271/sjweh.3830

Parent, M. E., Siemiatycki, J., Laplante, E., & Desy, M. (2003). Risk of lung cancer and mesothelioma associated with occupational exposure to asbestos: A population-based case-control study in Montreal, Canada. *Epidemiology, 14*(5), S69–S70. https://doi.org/10.1097/00001648-200309001-00156

Paz-Y-Mino, C., Sanchez, M. E., Arevalo, M., Munoz, M. J., Witte, T., De-La-Carrera, G. O., & Leone, P. E. (2007). Evaluation of DNA damage in an Ecuadorian population exposed to glyphosate. *Genetics and Molecular Biology, 30*(2), 456–460. https://doi.org/10.1590/s1415-47572007000300026

Pesch, B., Kendzia, B., Gustavsson, P., Jöckel, K. H., Johnen, G., Pohlabeln, H., et al. (2012). Cigarette smoking and lung cancerurelative risk estimates for the major histological types from a pooled analysis of case-control studies. *International Journal of Cancer, 131*(5), 1210–1219. https://doi.org/10.1002/ijc.27339

Portier, C. J. (2020). A comprehensive analysis of the animal carcinogenicity data for glyphosate from chronic exposure rodent carcinogenicity studies. *Environmental Health, 19*(1), 17. https://doi.org/10.1186/s12940-020-00574-1

Powell, H. A., Iyen-Omofoman, B., Hubbard, R. B., Baldwin, D. R., & Tata, L. J. (2013). The association between smoking quantity and lung cancer in men and women. *Chest, 143*(1), 123–129. https://doi.org/10.1378/chest.12-1068

Rahal, Z., El Nemr, S., Sinjab, A., Chami, H., Tfayli, A., & Kadara, H. (2017). Smoking and lung cancer: A geo-regional perspective. *Frontiers in Oncology, 7*, 7. https://doi.org/10.3389/fonc.2017.00194

Rezende, E. C. N., Carneiro, F. M., de Moraes, J. B., & Wastowski, I. J. (2021). Trends in science on glyphosate toxicity: A scientometric study. *Environmental Science and Pollution Research, 28*(40), 56432–56448. https://doi.org/10.1007/s11356-021-14556-4

Samet, J. M. (2015). The IARC monographs: Critics and controversy. *Carcinogenesis, 36*(7), 707–709. https://doi.org/10.1093/carcin/bgv062

Sapbamrer, R., Khacha-ananda, S., Sittitoon, N., Wunnapuk, K., Seesen, M., Sidthilaw, S., et al. (2019). A longitudinal follow-up study of oxidative stress and DNA damage among farmers exposed to pesticide mixtures. *Environmental Science and Pollution Research, 26*(13), 13185–13194. https://doi.org/10.1007/s11356-019-04650-z

Schinasi, L., & Leon, M. E. (2014). Non-Hodgkin lymphoma and occupational exposure to agricultural pesticide chemical groups and active ingredients: A systematic review and meta-analysis. *International Journal of Environmental Research and Public Health, 11*(4), 4449–4527. https://doi.org/10.3390/ijerph110404449

Schrenk, D. (2018). What is the meaning of 'A compound is carcinogenic?'. *Toxicology Reports, 5*, 504–511. https://doi.org/10.1016/j.toxrep.2018.04.002

Seralini, G. E., Clair, E., Mesnage, R., Gress, S., Defarge, N., Malatesta, M., et al. (2012). Long term toxicity of a Roundup herbicide and a Roundup-tolerant genetically modified maize (Retracted). *Food and Chemical Toxicology, 50*(11), 4221–4231. https://doi.org/10.1016/j.fct.2012.08.005

Sidthilaw, S., Sapbamrer, R., Pothirat, C., Wunnapuk, K., & Khacha-ananda, S. (2022). Effects of exposure to glyphosate on oxidative stress, inflammation, and lung function in maize farmers, Northern Thailand. *BMC Public Health, 22*(1), 10. https://doi.org/10.1186/s12889-022-13696-7

Singh, R., Shukla, A., Kaur, G., Girdhar, M., Malik, T., & Mohan, A. (2024). Systemic analysis of glyphosate impact on environment and human health. *ACS Omega, 9*(6), 6165–6183. https://doi.org/10.1021/acsomega.3c08080

Sosa, V., Moline, T., Somoza, R., Paciucci, R., Kondoh, H., & Lleonart, M. E. (2013). Oxidative stress and cancer: An overview. *Ageing Research Reviews, 12*(1), 376–390. https://doi.org/10.1016/j.arr.2012.10.004

Stapelfeld, C., Dammann, C., & Maser, E. (2020). Sex-specificity in lung cancer risk. *International Journal of Cancer, 146*(9), 2376–2382. https://doi.org/10.1002/ijc.32716

Steliga, M. A., & Dresler, C. M. (2011). Epidemiology of lung cancer: Smoking, secondhand smoke, and genetics. *Surgical Oncology Clinics of North America, 20*(4), 605-+. https://doi.org/10.1016/j.soc.2011.07.003

Talibov, M., Sormunen, J., Weiderpass, E., Kjaerheim, K., Martinsen, J. I., Sparen, P., et al. (2019). Workplace diesel exhausts and gasoline exposure and risk of colorectal cancer in four Nordic countries. *Safety and Health at Work, 10*(2), 141–150. https://doi.org/10.1016/j.shaw.2019.01.001

Tarazona, J. V., Court-Marques, D., Tiramani, M., Reich, H., Pfeil, R., Istace, F., & Crivellente, F. (2017). Glyphosate toxicity and carcinogenicity: A review of the scientific basis of the European Union assessment and its differences with IARC. *Archives of Toxicology, 91*(8), 2723–2743. https://doi.org/10.1007/s00204-017-1962-5

Torretta, V., Katsoyiannis, I. A., Viotti, P., & Rada, E. C. (2018). Critical review of the effects of glyphosate exposure to the environment and humans through the food supply chain. *Sustainability, 10*(4), 20. https://doi.org/10.3390/su10040950

Villeneuve, P. J., Parent, M. E., Sahni, V., Johnson, K. C., & Canadian Canc, R. (2011). Occupational exposure to diesel and gasoline emissions and lung cancer in Canadian men. *Environmental Research, 111*(5), 727–735. https://doi.org/10.1016/j.envres.2011.04.003

Ward, E. M. (2018). Glyphosate use and cancer incidence in the agricultural health study: An epidemiologic perspective. *Jnci-Journal of the National Cancer Institute, 110*(5), 446–447. https://doi.org/10.1093/jnci/djx247

Weisenburger, D. D. (2021). A review and update with perspective of evidence that the herbicide glyphosate (roundup) is a cause of non-Hodgkin lymphoma. *Clinical Lymphoma Myeloma & Leukemia, 21*(9), 621–630. https://doi.org/10.1016/j.clml.2021.04.009

West, G. H., Sokas, R. K., & Welch, L. S. (2019). Change in prevalence of asbestos-related disease among sheet metal workers 1986 to 2016. *American Journal of Industrial Medicine, 62*(7), 609–615. https://doi.org/10.1002/ajim.22998

Williams, G. M., Aardema, M., Acquavella, J., Berry, S. C., Brusick, D., Burns, M. M., et al. (2016). A review of the carcinogenic potential of glyphosate by four independent expert panels and comparison to the IARC assessment (Publication with Expression of Concern. See vol. 48, pg. 907, 2018). *Critical Reviews in Toxicology, 46*, 3–20. https://doi.org/10.1080/10408444.2016.1214677

Wolf, D. C., Cohen, S. M., Boobis, A. R., Dellarco, V. L., Fenner-Crisp, P. A., Moretto, A., et al. (2019). Chemical carcinogenicity revisited 1: A unified theory of carcinogenicity based on contemporary knowledge. *Regulatory Toxicology and Pharmacology, 103*, 86–92. https://doi.org/10.1016/j.yrtph.2019.01.021

Wozniak, E., Sicinska, P., Michalowicz, J., Wozniak, K., Reszka, E., Huras, B., et al. (2018). The mechanism of DNA damage induced by Roundup 360 PLUS, glyphosate and AMPA in human peripheral blood mononuclear cells – genotoxic risk assessement. *Food and Chemical Toxicology, 120*, 510–522. https://doi.org/10.1016/j.fct.2018.07.035

Wozniak, E., Reszka, E., Jablonska, E., Mokra, K., Balcerczyk, A., Huras, B., et al. (2020). The selected epigenetic effects of aminomethylphosphonic acid, a primary metabolite of glyphosate on human peripheral blood mononuclear cells (in vitro). *Toxicology In Vitro, 66*, 8. https://doi.org/10.1016/j.tiv.2020.104878

Xu, L., Qu, Y. H., Chu, X. D., Wang, R., Nelson, H. H., Gao, Y. T., & Yuan, J. M. (2015). Urinary levels of N-nitroso compounds in relation to risk of gastric cancer: Findings from the Shanghai cohort study. *PLoS One, 10*(2). https://doi.org/10.1371/journal.pone.0117326

Yamaguchi, N. (1999). Uncertainty in risk characterization of weak carcinogens. In A. J. Bailer, C. Maltoni, J. C. Bailar, F. Belpoggi, J. V. Brazier, & M. Soffritti (Eds.), *Uncertainty in the risk assessment of environmental and occupational hazards* (Vol. 895, pp. 338–347). New York Acad Sciences.

Zeng, J. F., Tang, Y. Y., Wu, P., Fang, X., Wang, W., Fan, Y. H., et al. (2019). Alcohol consumption, tobacco smoking, betel quid chewing and oral health associations with hypopharyngeal cancer among men in Central South China: A case-control study. *Cancer Management and Research, 11*, 6353–6364. https://doi.org/10.2147/cmar.S203439

Zhang, L., Rana, I., Shaffer, R. M., Taioli, E., & Sheppard, L. (2019). Exposure to glyphosate-based herbicides and risk for non-Hodgkin lymphoma: A meta-analysis and supporting evidence. *Mutation Research-Reviews in Mutation Research, 781*, 186–206. https://doi.org/10.1016/j.mrrev.2019.02.001

Zhou, W., Li, M. M., & Achal, V. (2025). A comprehensive review on environmental and human health impacts of chemical pesticide usage. *Emerging Contaminants, 11*(1), 13. https://doi.org/10.1016/j.emcon.2024.100410

Chapter 4
Glyphosate and Endocrinology

The main body functions are digestion (which delivers energy to keep the system running), excretion (getting rid of useless or toxic material/waste), circulation (pumping the blood through the whole organism hereby reaching all cells and delivering vital ingredients), respiration (using the oxygen of the air needed for the production of energy-containing ATP molecules), and other functions needed for survival, such as immune defense and reproduction. The brain controls all these functions in harmonious coexisting and concerted activities. The survival of the human species is dependent on how this machinery is maintained, and because of the complexity of the machine, the regulatory systems underscoring the functioning of the machine are equally complex and vulnerable to external and internal influences. If we induce alterations that disturb regulatory systems, we become ill. I will discuss two main control systems: In this chapter, I introduce the endocrine system; in the next chapter, the focus will be on the neuronal system.

The Human Endocrine System

The endocrine system in humans consists of several hormone-secreting glands distributed throughout the whole body. These are in the brain: the pituitary gland and the hypothalamus. The pituitary gland secretes growth hormone, thyroid-stimulating hormone, prolactin, follicle-stimulating hormone, and others; it is considered the main endocrine organ that regulates the activity of other endocrine glands. The hypothalamus functions as a feedback control mechanism of the pituitary gland and is also involved in the interaction between the endocrine system and the nervous system. The organs outside the brain include the parathyroid glands, which release the parathyroid hormone regulating calcium balance; the thyroid gland, which releases thyroxine (T4) and triiodothyronine (T3), which regulate metabolism; the adrenal glands, which produce important hormones such as adrenaline in the

N. Fraeyman, *Glyphosate 2023–2033*,
https://doi.org/10.1007/978-3-031-97540-0_4

medulla and cortisol and other steroids in the cortex; the pancreas, which secretes insulin that regulates the blood sugar balance; the ovaries in females, which produce estrogen and are regulatory for reproductive functions; and the testes in males, which produce testosterone and hormones that are regulatory for sperm production. All hormones are secreted by the glands into the extracellular space and are ultimately transported through the whole body by the blood. A simplified presentation is given at https://www.nursingtimes.net/primary-care/endocrine-system-1-overview-of-the-endocrine-system-and-hormones-26-04-2021/. Many textbooks of variable degrees of detail are available, but this is not required to understand the remainder of the chapter. Notably, this is a very complex system in which the quality of forward and backward control keeps the whole system in balance or homeostasis.

The organs and hormones are shown in Fig. 4.1.

The second important piece of information related to endocrine (and neurological—see next chapter) hormones is the mechanism by which they exert their activity. Hormones, such as insulin, thyroid hormones, and steroids, are message-carrying molecules conveniently called "first messengers." They dictate the functions of other organs, including other endocrine glands, both in normal, basic conditions and in stimulated or uncontrolled situations. The message of some hormones is conveyed to the cells of the target organ through receptors, i.e., specialized cellular structures in the cell membranes that are capable of capturing the message-carrying

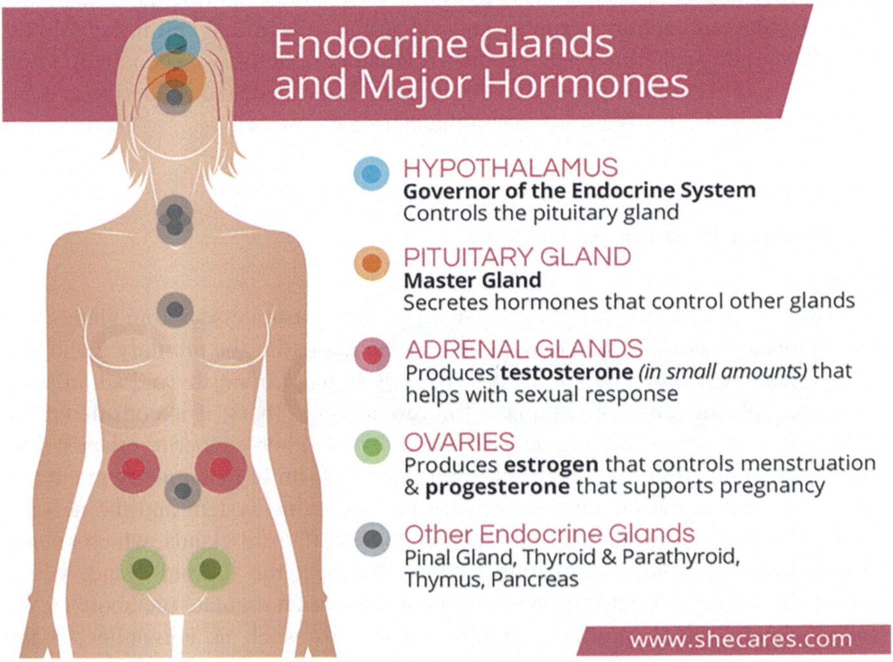

Fig. 4.1 Schematic representation of the most important endocrine glands

molecule and translating the message compatible with the properties of the target organ that provides the appropriate response. For peptide-like hormones such as insulin, these receptors are outside the target cell, and the message must be converted inside the cell into a "second message" that prompts the cell to respond to the first message. The group of steroid message molecules/hormones has a similar system, but the receptor for the steroids is inside the cell. Therefore, steroids must pass through the cell membrane; the remainder of the process is comparable. For steroids, the cellular response occurs at the level of the nucleus, including its influence on DNA transcription and subsequent protein synthesis.

The third biological mechanism involves the transduction of the first signal conventionally called the "signal transduction system." For the hormones that do not enter the cell, this mechanism has to "translate" the message, which stays outside of the cell in such a way that the target cell understands the message and can respond in the way the message requires. This signal transduction system is an interplay of interfering intracellular enzymes that are, as a rule in biological processes, fine-tuned and under strict control. The full description of this process is beyond the scope of this book; suffice to refer to some recent reviews (Calebiro, 2021; Cerione, 2018; Komolov & Benovic, 2018).

Finally, the target cell, eventually another endocrine gland, will respond. The cellular reaction depends on the cell type: A muscle cell will contract or relax, a nerve cell will fire a stimulus, insulin will induce glucose uptake in cells along with other effects, and endocrine glands will secrete their hormones.

These three-step mechanisms are depicted in Fig. 4.2.

Evidence for the Endocrine-Disrupting Properties of Pesticides Including Glyphosate

Endocrine disruption by chemicals is a disputable issue given the complexity of the endocrine system, the multiple feedback systems, and the influence of one interaction at a particular target on the rest of the system. In 2020, a system for organizing and interpreting relevant literature on the subject was presented (La Merrill et al., 2020). The authors identified ten key characteristics (KCs), which eventually turn a chemical into an endocrine-disrupting molecule. According to Muñoz et al. (2021), glyphosate conforms to eight out of the ten KCs.

The toxicity of some pesticides to the endocrine system has been studied in detail. Atrazine, which was banned from use in the EU in 2004, has clear endocrine-disrupting properties, and its molecular mechanisms have been established (Zhao et al., 2024). A similar conclusion is drawn for the other agrochemicals endosulfan and DDT (Vandenberg et al., 2020). Warner et al. (2020) described the modes of action of endocrine-disrupting molecules at different targets within the endocrine system, including the role of glyphosate at the hypothalamic–pituitary–adrenal axis and altered levels of progesterone and oestradiol. In in vivo experiments,

https://pressbooks.atlanticoer-relatlantique.ca/advcellbiol/chapter/56-signalling-across-the-plasma-membrane/

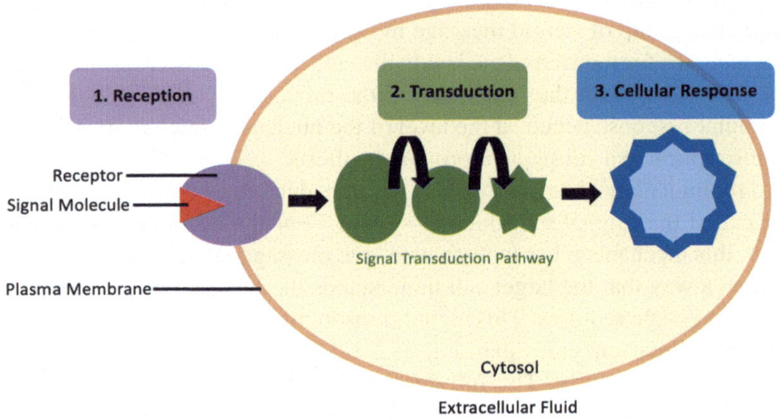

https://ib.bioninja.com.au/signal-transduction/

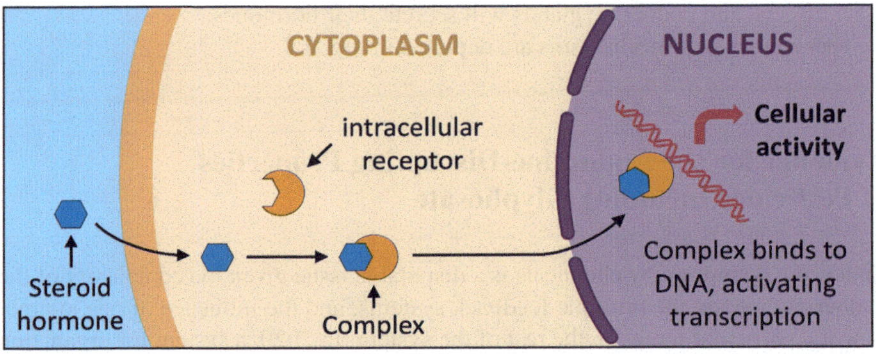

Fig. 4.2 Summary of the main steps of the signal transduction system. https://quizlet.com/244360371/signal-transduction-pathway-flash-cards/. https://old-ib.bioninja.com.au/options/option-d-human-physiology/d5-hormones-and-metabolism/types-of-hormones.html

glyphosate-based herbicides were shown to alter female fertility in animals (Ingaramo et al., 2020; Milesi et al., 2020), indicating interactions with the reproductive system. Lesseur et al. (2021, 2022) reported that the gestational length was related to the presence of glyphosate in the urine using data from a small cohort of women with 69 preterm births and 94 term births. Similarly, in a limited US Central Indiana cohort of 71 women, more than 90% had glyphosate in their urine (3.4 ng/ml on average). A significant ($p < 0.02$) correlation between gestational length and urinary glyphosate concentration was established (Parvez et al., 2018). Despite these results, coupling gestational length unequivocal to glyphosate remains disputable. An overview of endocrine-disrupting chemicals and health effects is given by L. Vandenberg (https://www.health.belgium.be/sites/default/files/uploads/fields/fpshealth_theme_file/general_meeting_2019_shc_prof._vandenberg.pdf) and by G. Schoeters (https://www.health.belgium.be/sites/default/files/uploads/fields/fpshealth_theme_file/generel_meeting_2019_shc_prof._schoeters.pdf).

The Superior Health Council (Belgium) is very concerned about the exposure to chemicals with mutagenic and endocrine-disrupting properties. In 2019, the SHC published an advice on this point, particularly related to exposure early in life (SCH—https://www.hgr-css.be/file/download/494a5627-760b-4940-8d90-bc7c96cf64c7/vUvfvFqG0f3MrUOTMdRuhRwLNbIEvX7dfGtkgEZCI3d.pdf).

Glyphosate has estrogen receptor properties and downregulates aromatase enzymatic activity at glyphosate concentrations lower than those used for agricultural purposes (Milesi et al., 2021). Aromatase is a key enzyme and a critical step in estrogen biosynthesis; reducing enzymatic activity hampers gametogenesis, sex differentiation, and reproduction.

Even more concerning is the possibility that epigenetic alterations become heritable and become visible in the second and subsequent generations (Kubsad et al., 2019). This transgenerational effect has been proven beyond any doubt for pesticides such as atrazine and DDT (Van Cauwenbergh et al., 2020), and there are reasons to assume that glyphosate behaves similarly. The glyphosate-induced changes in the endocrine system are visible in the offspring of exposed dams.

Other results have shown that glyphosate-based herbicide formulations, not glyphosate alone, affect adult rat testes (Johansson et al., 2018). Notably, in these experiments, the rats were dosed for only 2 weeks via oral gavage, which is a short period and an abnormal method of exposure. The interpretation of these results is difficult.

Most of these studies are performed with animals, and virtually no data are available on the effects of glyphosate on, for example, human fertility, both male and female. Notably, glyphosate is found in human sperm in the French population and is linked to increased oxidative stress (Vasseur et al., 2024). This conforms not only with the finding that glyphosate is present everywhere in the environment, including in all compartments of humans, but it also opens the possibility of a direct influence of glyphosate on the properties of spermatozoa. An effect on mitochondrial respiration (Ferramosca et al., 2021) and sperm motility (Anifandis et al., 2018) has been confirmed. The ability of reproductive toxicants to alter human fecundity quasi-irreversibly is difficult to predict but is at least worrying (Aitken, 2024).

The second remark concerns the levels of exposure. The low concentrations of chemicals in the blood, urine, or other body fluids are frequently used as an argument in favor of their further use since the concentrations did not reach the levels set by regulating governmental authorities. However, particularly related to the endocrine (and neuronal—see next chapter) system, the blood concentrations of hormones are extremely low in the picogram and sub-picogram ranges, and these are the concentrations at which they are active via dedicated receptors. The concentrations of environmentally harmful chemicals are equally low and thus can compete with bodily chemicals for receptors, as shown for glyphosate-based herbicides and the retinoic acid signaling system (Paganelli et al., 2010). Even if not all biomechanical details are known, this might lead to serious deformations in normal situations.

The European Food Safety Authority (EFSA)EFSA's point of view on endocrine disruption, in general, has been published several times, and an overview of the most important documents is given in Chap. 11, Annex 3.

The EU has identified a number of endocrine-disrupting molecules in three lists (https://edlists.org/the-ed-lists). List I contains identified disruptors, List II shows molecules under evaluation, and List III shows endocrine disruptors considered by national authorities. The EFSA takes as the basis for its decisions the criteria laid down in 2013 and further detailed in 2018 (see the EFSA websites). One of these criteria is the causal relationship between an endocrine biological effect of a substance (for which experimental evidence is available in many cases) and the overt bodily expression of any effect as the consequence of a disturbed endocrine equilibrium. The applicability of this criterion is questionable because of the biological complexity of the endocrine system. It will be very difficult and, in many cases, impossible to prove this causal relationship.

The glyphosate molecule is a good example of the difficulty of applying the EFSA criteria. Glyphosate at low concentrations is an aromatase inhibitor, which is a key enzyme in estrogenic endocrinology (i.e., "experimental evidence"). The data on possible effects in humans are limited for evident reasons, but in experimental animals, decreased female fertility is a phenotype of endocrine disturbance (i.e., "overt bodily expression in animals"), although, in humans, other factors might influence fertility. From this, EFSA concluded that glyphosate has no estrogenic properties. However, estrogenic hormones are one type of hormone, and other types such as peptide-type hormones (insulin, oxytocin, melatonin, etc.) or small molecules with hormonal functions (serotonin and thyroid hormones) are not considered.

As mentioned many times before, we wait until scientific evidence emerges, but until then, we are all exposed to potentially endocrine-disrupting molecules, among which glyphosate is a prominent representative because of its ubiquitous use. The application of the precautionary principle is indicated. Notably, in the key legal documents of the EU concerning endocrine disruption (regulation EU 528/2012 and EC 1107/2009), the word "precautionary principle" is not mentioned once.

Taken together, interest in endocrine-disrupting substances within the EU is high as far as legal measures are concerned (Duh-Leong et al., 2023; Kassotis et al., 2020; Liszewska et al., 2024). It remains, however, questionable how far these legal constraints have resulted in concrete measures in the real world. Despite the impressive number of suspected endocrine molecules in the EFSA lists, only eight pesticides are banned because of their endocrine-disrupting properties. There are reasons to believe that precaution was not the main concern when judging glyphosate for its potential endocrine-disrupting properties.

References

Aitken, R. J. (2024). The global decline in human fertility: The post-transition trap hypothesis. *Life-Basel, 14*(3), 19. https://doi.org/10.3390/life14030369

Anifandis, G., Amiridis, G., Dafopoulos, K., Daponte, A., Dovolou, E., Gavriil, E., et al. (2018). The in vitro impact of the herbicide roundup on human sperm motility and sperm mitochondria. *Toxics, 6*(1). https://doi.org/10.3390/toxics6010002

Calebiro, D. (2021). Signalling by G protein-coupled receptors: Why space and time matter. *European Journal of Endocrinology, 184*(2), R41–R49. https://doi.org/10.1530/EJE-20-0890

Cerione, R. A. (2018). The experiences of a biochemist in the evolving world of G protein-dependent signaling. *Cellular Signalling, 41*, 2–8. https://doi.org/10.1016/j.cellsig.2017.02.016

Duh-Leong, C., Maffini, M. V., Kassotis, C. D., Vandenberg, L. N., & Trasande, L. (2023). The regulation of endocrine-disrupting chemicals to minimize their impact on health. *Nature Reviews Endocrinology, 15*. https://doi.org/10.1038/s41574-023-00872-x

Ferramosca, A., Lorenzetti, S., Di Giacomo, M., Murrieri, F., Coppola, L., & Zara, V. (2021). Herbicides glyphosate and glufosinate ammonium negatively affect human sperm mitochondria respiration efficiency. *Reproductive Toxicology, 99*, 48–55. https://doi.org/10.1016/j.reprotox.2020.11.011

Ingaramo, P., Alarcón, R., Muñoz-de-Toro, M., & Luque, E. H. (2020). Are glyphosate and glyphosate-based herbicides endocrine disruptors that alter female fertility? *Molecular and Cellular Endocrinology, 518*, 12. https://doi.org/10.1016/j.mce.2020.110934

Johansson, H. K. L., Schwartz, C. L., Nielsen, L. N., Boberg, J., Vinggaard, A. M., Bahl, M. I., & Svingen, T. (2018). Exposure to a glyphosate-based herbicide formulation, but not glyphosate alone, has only minor effects on adult rat testis. *Reproductive Toxicology, 82*, 25–31. https://doi.org/10.1016/j.reprotox.2018.09.008

Kassotis, C. D., Vandenberg, L. N., Demeneix, B. A., Porta, M., Slama, R., & Trasande, L. (2020). Endocrine-disrupting chemicals: Economic, regulatory, and policy implications. *Lancet Diabetes & Endocrinology, 8*(8), 719–730. Retrieved from <Go to ISI>://WOS:000551458500007.

Komolov, K. E., & Benovic, J. L. (2018). G protein-coupled receptor kinases: Past, present and future. *Cellular Signalling, 41*, 17–24. https://doi.org/10.1016/j.cellsig.2017.07.004

Kubsad, D., Nilsson, E. E., King, S. E., Sadler-Riggleman, I., Beck, D., & Skinner, M. K. (2019). Assessment of glyphosate induced epigenetic transgenerational inheritance of pathologies and sperm epimutations: Generational toxicology. *Scientific Reports, 9*. https://doi.org/10.1038/s41598-019-42860.0

La Merrill, M. A., Vandenberg, L. N., Smith, M. T., Goodson, W., Browne, P., Patisaul, H. B., et al. (2020). Consensus on the key characteristics of endocrine-disrupting chemicals as a basis for hazard identification. *Nature Reviews Endocrinology, 16*(1), 45–57. https://doi.org/10.1038/s41574-019-0273-8

Lesseur, C., Pirrotte, P., Pathak, K. V., Manservisi, F., Mandrioli, D., Belpoggi, F., et al. (2021). Maternal urinary levels of glyphosate during pregnancy and anogenital distance in newborns in a US multicenter pregnancy cohort. *Environmental Pollution, 280*. https://doi.org/10.1016/j.envpol.2021.117002

Lesseur, C., Pathak, K., Pirrotte, P., Martinez, M. N., Ferguson, K. K., Barrett, E. S., et al. (2022). Urinary glyphosate concentration in pregnant women in relation to length of gestation. *Environmental Research, 203*, 7. https://doi.org/10.1016/j.envres.2021.111811

Liszewska, M., Czaja, K., Korcz, W., Lewinski, R., & Strucinski, P. (2024). Endocrine-disrupting chemicals – pesticide regulatory issues from the EU perspective. *Regulatory Toxicology and Pharmacology, 154*, 9. https://doi.org/10.1016/j.yrtph.2024.105735

Milesi, M. M., Lorenz, V., Beldomenico, P. M., Vaira, S., Varayoud, J., & Luque, E. H. (2020). Response to comments on: Perinatal exposure to a glyphosate-based herbicide impairs female reproductive outcomes and induces second-generation adverse effects in Wistar rats

(vol 93, pg 3635, 2019). *Archives of Toxicology, 94*(8), 2897–2898. https://doi.org/10.1007/s00204-020-02814-2
Milesi, M. M., Lorenz, V., Durando, M., Rossetti, M. F., & Varayoud, J. (2021). Glyphosate herbicide: Reproductive outcomes and multigenerational effects. *Frontiers in Endocrinology, 12*, 22. https://doi.org/10.3389/fendo.2021.672532
Muñoz, J. P., Bleak, T. C., & Calaf, G. M. (2021). Glyphosate and the key characteristics of an endocrine disruptor: A review. *Chemosphere, 270*, 15. https://doi.org/10.1016/j.chemosphere.2020.128619
Paganelli, A., Gnazzo, V., Acosta, H., López, S. L., & Carrasco, A. E. (2010). Glyphosate-based herbicides produce teratogenic effects on vertebrates by impairing retinoic acid signaling. *Chemical Research in Toxicology, 23*(10), 1586–1595. https://doi.org/10.1021/tx1001749
Parvez, S., Gerona, R. R., Proctor, C., Friesen, M., Ashby, J. L., Reiter, J. L., et al. (2018). Glyphosate exposure in pregnancy and shortened gestational length: A prospective Indiana birth cohort study. *Environmental Health, 17*, 12. https://doi.org/10.1186/s12940-018-0367-0
Van Cauwenbergh, O., Di Serafino, A., Tytgat, J., & Soubry, A. (2020). Transgenerational epigenetic effects from male exposure to endocrine-disrupting compounds: A systematic review on research in mammals. *Clinical Epigenetics, 12*(1), 23. https://doi.org/10.1186/s13148-020-00845-1
Vandenberg, L. N., Najmi, A., & Mogus, J. P. (2020). Agrochemicals with estrogenic endocrine disrupting properties: Lessons learned? *Molecular and Cellular Endocrinology, 518*, 13. https://doi.org/10.1016/j.mce.2020.110860
Vasseur, C., Serra, L., El Balkhi, S., Lefort, G., Rame, C., Froment, P., & Dupont, J. (2024). Glyphosate presence in human sperm: First report and positive correlation with oxidative stress in an infertile French population. *Ecotoxicology and Environmental Safety, 278*, 11. https://doi.org/10.1016/j.ecoenv.2024.116410
Warner, G. R., Mourikes, V. E., Neff, A. M., Brehm, E., & Flaws, J. A. (2020). Mechanisms of action of agrochemicals acting as endocrine disrupting chemicals. *Molecular and Cellular Endocrinology, 502*, 13. https://doi.org/10.1016/j.mce.2019.110680
Zhao, H. T., Qian, H. H., Cui, J. W., Ge, Z. L., Shi, J. J., Huo, Y. C., et al. (2024). Endocrine toxicity of atrazine and its underlying mechanisms. *Toxicology, 505*, 14. https://doi.org/10.1016/j.tox.2024.153846

Chapter 5
Glyphosate and the Neuronal System

This chapter focuses on the influence of glyphosate and pesticides on the neuronal system. There is much in common with the previous chapter on the endocrine system.

Basics of the Neuronal System

In the previous chapter on the endocrine system, the concepts of hormones, receptors, signal transduction, and cellular response were explained. Similar biological systems are known for the human neuronal system; signaling molecules are called neurotransmitters.

First, we need to distinguish between the central nervous system, which is the brain and the spinal cord, and the peripheral nervous system outside the brain in the remainder of the body. Both are intimately linked, and both are important when discussing the possible impact of glyphosate.

Second, the basic unit of both the central and the peripheral neuronal system is the neuron. These are irregular-shaped cells with thread-like branches called dendrites or axons; signals from neurons to other neurons in the brain or to organs in the peripheral system are transported via these axons. Different types of neurons are known, such as motor neurons, which originate in the spinal cord and control all the organs in the body; sensory neurons, which are located in sensory organs (e.g., skin) and send messages to the brain via interneurons; and cranial neurons, which are located inside the brain.

Essentially, two parallel neuronal systems are known: the adrenergic or sympathetic or ortho-sympathetic nervous system, with the catecholamines adrenaline (or epinephrine) and noradrenaline (or norepinephrine) as major neurotransmitters, and the cholinergic system or the parasympathetic nervous system, with acetylcholine as the major neurotransmitter. The two systems together are called the autonomic

N. Fraeyman, *Glyphosate 2023–2033*,
https://doi.org/10.1007/978-3-031-97540-0_5

nervous system. It is "autonomic" because it regulates involuntary actions in the body, such as respiration, heart rate, and the tonus of blood vessels. Both the sympathetic and parasympathetic neuronal systems originate in the brain and affect organs via motor neurons. This is shown in Fig. 5.1.

As a rule, the sympathetic and parasympathetic neuronal systems have opposite effects on organs: stimulatory or inhibitory; normal and optimal functioning of organs depends on the interplay between the adrenergic and cholinergic systems.

Like in the endocrine system, neurotransmitters are called first messengers. The receptors involved in the neuronal system are multiple and have been studied extensively; the principle of a receptor, second messenger, and signal transduction are explained in the previous chapter. The receptors of the sympathetic system are named alpha and beta, and several subtypes have been described for each. The corresponding receptors of the parasympathetic system are the muscarinic and nicotinic receptors. The signal transduction system of the sympathetic nervous system has been the subject of the Nobel Prize in 2012 and involves a cascade of enzymatic activities leading to the cellular response. The full description of neurotransmitters and receptors is beyond the scope of this book.

Interfering with neurotransmitters at the level of their receptors is an appealing possibility for the development of medical drugs. For example, one possible pharmacological treatment for hypertension is lowering the heart's pumping force and/

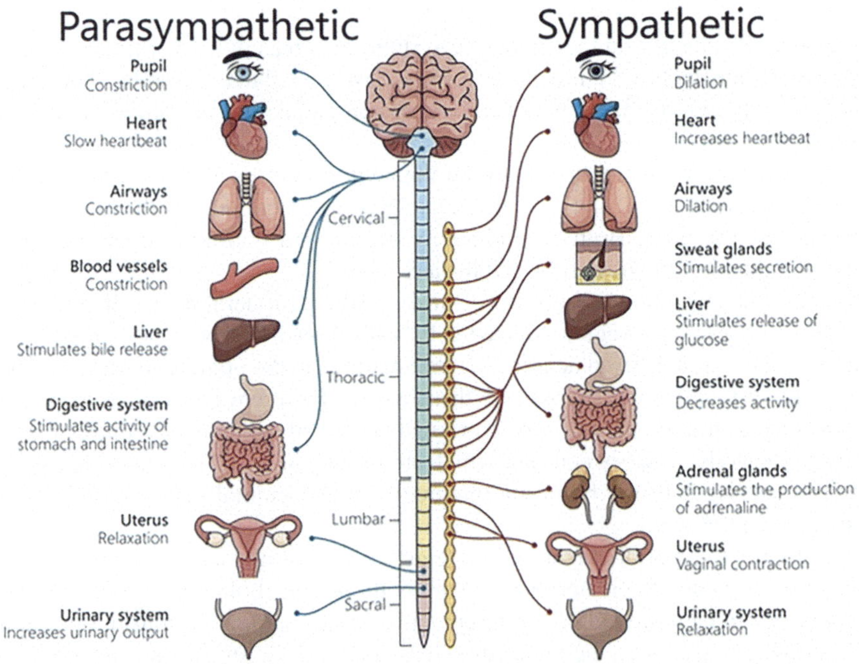

Fig. 5.1 Schematic representation of the sympathetic and parasympathetic neuronal system and their target organs

or rate by inhibiting the sympathetic system's stimulatory effect via the use of an inhibitor molecule that prevents the stimulation of heart muscle cells by catecholamines.

Finally, neurotransmitters are very powerful molecules that can jeopardize the body's normal functions. Excess noradrenaline will increase the pumping rate of the heart and increase blood pressure. This might harm other organs in the body, not in the least the delicate small blood veins that need to cope with increased blood pressure. Therefore, the body has powerful deactivation systems at hand that can eliminate the activity of the transmitters by enzymatic degradation. These enzymes are for catecholamine monoamine oxidase and catechol-orthomethyltransferase; for acetylcholine transmitters, the enzyme is acetylcholinesterase. Some insecticides, such as parathion and malathion, are potent inhibitors of acetylcholinesterase.

From this and the previous short introduction, the reader should be convinced of the complexity of cellular control systems and—more than anything else—be convinced of the beauty of the functioning of our body due to a perfect balance between all influences. We take it for granted and consider it normal even if we do not understand all the details. If, however, the balance is distorted, the vulnerability of the system becomes evident, and we become sick. We must treasure this delicate system at the individual and societal levels. The possibility of disturbing our delicate bodily functioning must be kept as low as possible.

Pesticides and Glyphosate; Interactions and Neuronal Diseases

There are multiple animal experiments showing the direct activity of glyphosate in the brain (Martínez et al., 2018; Gallegos et al., 2020; Luna et al., 2021; Moser et al. 2022a, b; Costas-Ferreira et al., 2023; Izumi et al., 2023; Ojiro et al., 2023; Chávez-Reyes et al., 2024a; Costas-Ferreira et al., 2024; Chiantia et al., 2025). Despite these convincing results, the extrapolation of these findings to the human situation remains unclear. The administration of the compound in rodents is often far from human-like exposure (e.g., glyphosate is administered through the nose of the mouse). Hence, a cautious interpretation of these results is needed.

Observations in humans have revealed several neurodevelopmental and behavioral effects upon exposure to different pesticides, and the exposure to pesticides itself has been confirmed by longitudinal measurements of biomarkers in blood and urine (Etiemble & Cordier, 2022; Zhou et al., 2025). Delayed developmental effects, impairment of fine motor skills and visual function, and an increased risk of autism spectrum disorders are associated with glyphosate exposure (Madani & Carpenter, 2022). Exposure during pregnancy leads to enhanced expression of autism in offspring (He et al., 2022).

The most astonishing report (especially for the media) was probably that of Samsel and Seneff (2013), who claimed that glyphosate inhibits the cytochrome

P450 detoxification pathways in the liver and substitutes glycine in polypeptides. In particular, Seneff suggested that glyphosate would lead to 50% of children having autism by 2030 (Samsel & Seneff, 2013).

Samsel and Seneff's article was critically examined, and the conclusions were straightforward: P450 inhibition was absent, and the enzyme was even found to be slightly activated. The chelation of manganese and its link with neurological diseases was not investigated, but the supposed replacement of glycine with glyphosate was wrong (Mesnage & Antoniou, 2017, 2018). These authors state in the abstract: "*We found that [Samsel and Seneff] inappropriately employ a deductive reasoning approach on the basis of syllogism. We found that their conclusions are not supported by the available scientific evidence. Thus, the mechanisms and vast range of conditions proposed to result from glyphosate toxicity presented by Samsel and Seneff in their commentaries are at best unsubstantiated theories, speculations, or simply incorrect.*" In addition, the possibility that glyphosate mimics the amino acid glycine in various enzymatic reactions was proven to be wrong (Antoniou et al., 2019). To use the famous words of Wolfgang Pauli (Nobel Prize in Physics 1945): "This isn't right, it's not even wrong."

Although the paper of Samsel and Seneff is inadequate, the impact of the environment on the development of autism spectrum disorder (ASD) should not be ignored (Sealey et al., 2016; von Ehrenstein et al., 2019). The role of endocrine-disrupting agents in the etiology of ASD has been reviewed (Moosa et al., 2018), demonstrating that, in addition to the genetic causes of ASD, the role of environmental chemical exposure is suggestive. These authors concluded that "*exposure to this class of chemicals can lead to persistent changes in gene expression and phenotype, which may in turn contribute to transgenerational inheritance of autism spectrum disorder.*" The role of environmental factors in the development of ASD is highly probable (Sealey et al., 2016).

The second disease at the center of discussion is Parkinson's disease. The relationship between pesticide exposure and brain damage leading to Parkinson's disease has been reviewed recently (Bloem & Boonstra, 2023; Bloem et al., 2024). In France, the disease is considered an occupational disease, and compensation policies for people working in agriculture following years of exposure have been foreseen since 7/5/2013. Epidemiological research in France is extensive and convincing (Elbaz & Moisan, 2016; Kab et al. 2017a, b; Wanneveich et al., 2018; Perrin et al., 2019, 2020). In other countries (e.g., Belgium, Germany, the Netherlands, Sweden, and Italy), Parkinson's disease is not formally considered an occupational disease in the legal sense of the word but exposure to pesticides during agricultural work is considered a risk factor for developing Parkinson's disease. In these countries, patients are entitled to support if a causal relationship between work and disease is feasible.

Taken together, the neurotoxicity of pesticides in general and glyphosate in particular is well documented, both from fundamental basic biomedical research with animals and from in vitro experiments. The results from epidemiological studies are similarly convincing, leading to the recognition of Parkinson's disease as an occupational disease. As this is possible in France, the question is why this is not the case in other European countries.

Mechanistic Considerations on the Neurotoxicity of Glyphosate

The complexity of the neuronal system implies many possible targets for interaction with glyphosate, which has been confirmed in several recent overview papers (Richardson et al., 2019; Costas-Ferreira et al., 2022; Chang et al., 2023; Bloem et al., 2024; Fan et al., 2024).

Several experimental studies are available on the inhibition of acetylcholinesterase in the blood by glyphosate, either directly or indirectly through an effect of glyphosate on distant target organs, releasing inhibitory molecules for the enzyme (Chávez-Reyes et al., 2024b) and in the brains of rats (Costas-Ferreira et al., 2023; Chávez-Reyes et al., 2024a). In contrast, no inhibition of human acetylcholinesterase was found, questioning the value of the animal experimental results (Kolic et al., 2024).

An action of glyphosate on brain acetylcholinesterase has also been described (Chávez-Reyes et al., 2024a), although the central GABA-aminergic system might be affected (Cattani et al., 2014; Martínez et al., 2018; Gallegos et al., 2020; Luna et al., 2021; Ojiro et al., 2023; Costas-Ferreira et al., 2023, 2024).

A recent paper described how glyphosate enters the rat brain, and a significant correlation was found between the concentrations of central glyphosate and TNF-alpha, which is a cellular proinflammatory cytokine (Winstone et al., 2024). These findings indicate that the effect of glyphosate is possibly second to the induction of inflammation (Izumi et al., 2023, 2024), such as the effect of glyphosate in the case of cancer, and that glyphosate is likely an endocrine-disrupting molecule via the production of reactive oxygen species.

Taken together, this field is in full development, and relevant new results are expected. The evidence suggests an interaction between glyphosate and the (human) neuronal system, whether direct or indirect. Basic biomedical details are still lacking, but the point is what kind of scientific results we need to be convinced of a neuronal effect of glyphosate. The importance is hard to underestimate in case the evidence points to a cause–effect link with neuronal diseases such as autism spectrum disorders or Parkinson's disease.

References

Antoniou, M. N., Nicolas, A., Mesnage, R., Biserni, M., Rao, F. V., & Martin, C. V. (2019). Glyphosate does not substitute for glycine in proteins of actively dividing mammalian cells. *BMC Research Notes, 12*(1), 7.

Bloem, B. R., & Boonstra, T. A. (2023). The inadequacy of current pesticide regulations for protecting brain health: The case of glyphosate and Parkinson's disease. *Lancet Planetary Health, 7*(12), E948–E949.

Bloem, B. R., Boonstra, T. A., Elbaz, A., & Vermeulen, R. C. H. (2024). Glyphosate and neurotoxicity – A call for scientific renewal. *Nature Reviews Neurology, 2*, 131–132.

Cattani, D., Cavalli, V., Rieg, C. E. H., Domingues, J. T., Dal-Cim, T., Tasca, C. I., Silva, F., & Zamoner, A. (2014). Mechanisms underlying the neurotoxicity induced by glyphosate-based herbicide in immature rat hippocampus: Involvement of glutamate excitotoxicity. *Toxicology, 320*, 34–45.

Chang, E. T., Odo, N. U., & Acquavella, J. F. (2023). Systematic literature review of the epidemiology of glyphosate and neurological outcomes. *International Archives of Occupational and Environmental Health, 96*(1), 1–26.

Chávez-Reyes, J., López-Lariz, C. H., & Marichal-Cancino, B. A. (2024a). Both acute glyphosate and the aminomethylphosphonic acid intoxication decreased the acetylcholinesterase activity in rat hippocampus, prefrontal cortex and gastrocnemius muscle. *Drug and Chemical Toxicology, 47*, 1033–1037.

Chávez-Reyes, J., Sarãchaga-Terrazas, F., Colis-Arenas, O. A., López-Lariz, C. H., Villalón, C. M., & Marichal-Cancino, B. A. (2024b). Aminomethylphosphonic acid (AMPA), a glyphosate metabolite, decreases plasma cholinesterase activity in rats. *Journal of Xenobiotics, 14*(2), 604–612.

Chiantia, G., Comai, D., Hidisoglu, E., Gurgone, A., Franchino, C., Carabelli, V., Marcantoni, A., & Giustetto, M. (2025). Glyphosate impairs both structure and function of GABAergic synapses in hippocampal neurons. *Neuropharmacology, 262*, 11.

Costas-Ferreira, C., Durán, R., & Faro, L. R. F. (2022). Toxic effects of glyphosate on the nervous system: A systematic review. *International Journal of Molecular Sciences, 23*(9), 35.

Costas-Ferreira, C., Duran, R., & Faro, L. F. (2023). Neurotoxic effects of exposure to glyphosate in rat striatum: Effects and mechanisms of action on dopaminergic neurotransmission. *Pesticide Biochemistry and Physiology, 193*, 14.

Costas-Ferreira, C., Durán, R., & Faro, L. R. F. (2024). Evaluation of the potential role of glutamatergic, cholinergic, and nitrergic systems in the dopamine release induced by the pesticide glyphosate in rat striatum. *Journal of Applied Toxicology, 44*(10), 1489–1503.

Elbaz, A., & Moisan, F. (2016). The scientific bases to consider Parkinson's disease an occupational disease in agriculture professionals exposed to pesticides in France. *Journal of Epidemiology and Community Health, 70*(4), 319–321.

Etiemble, J., & Cordier, S. (2022). Pesticides and neurodevelopment in children. *Environnement Risques & Sante, 21*(1), 51–65.

Fan, R. T., Zheng, Y. F., Naga, L. D. Y., Beeraka, N. M., Ramachandrappa, H. V. P., Suhail, S. M., Manjula, S. N., Bannimath, N., Sinelniknov, M. Y., & Liu, J. Q. (2024). Updated intrinsic role of phytochemicals against glyphosate-induced neurotoxicity: Systematic review. *Current Medicinal Chemistry, 32*, 3243–3257.

Gallegos, C. E., Bartos, M., Gumilar, F., Raisman-Vozari, R., Minetti, A., & Baier, C. J. (2020). Intranasal glyphosate-based herbicide administration alters the redox balance and the cholinergic system in the mouse brain. *Neurotoxicology, 77*, 205–215.

He, X., Tu, Y., Song, Y. W., Yang, G. H., & You, M. D. (2022). The relationship between pesticide exposure during critical neurodevelopment and autism spectrum disorder: A narrative review. *Environmental Research, 203*, 12.

Izumi, Y., O'Dell, K. A., & Zorumski, C. F. (2023). The herbicide glyphosate inhibits hippocampal long-term potentiation and learning through activation of pro-inflammatory signaling. *Scientific Reports, 13*(1), 12.

Izumi, Y., O'Dell, K. A., & Zorumski, C. F. (2024). Glyphosate as a direct or indirect activator of pro-inflammatory signaling and cognitive impairment. *Neural Regeneration Research, 19*(10), 2212–2218.

Kab, S., Moisan, F., & Elbaz, A. (2017a). Farming and incidence of motor neuron disease: French nationwide study. *European Journal of Neurology, 24*(9), 1191–1195.

Kab, S., Spinosi, J., Chaperon, L., Dugravot, A., Singh-Manoux, A., Moisan, F., & Elbaz, A. (2017b). Agricultural activities and the incidence of Parkinson's disease in the general French population. *European Journal of Epidemiology, 32*(3), 203–216.

Kolic, D., Pehar, V., & Kovarik, Z. (2024). Environmental exposure to glyphosate does not inhibit human acetylcholinesterase and butyrylcholinesterase. *Arhiv Za Higijenu Rada I Toksikologiju-Archives of Industrial Hygiene and Toxicology, 75*(1), 76–80.

Luna, S., Neila, L. P., Vena, R., Borgatello, C., & Rosso, S. B. (2021). Glyphosate exposure induces synaptic impairment in hippocampal neurons and cognitive deficits in developing rats. *Archives of Toxicology, 95*(6), 2137–2150.

Madani, N. A., & Carpenter, D. O. (2022). Effects of glyphosate and glyphosate-based herbicides like RoundupTM on the mammalian nervous system: A review. *Environmental Research, 214*, 9.

Martínez, M. A., Ares, I., Rodríguez, J. L., Martínez, M., Martínez-Larrañaga, M. R., & Anadón, A. (2018). Neurotransmitter changes in rat brain regions following glyphosate exposure. *Environmental Research, 161*, 212–219.

Mesnage, R., & Antoniou, M. N. (2017). Facts and fallacies in the debate on glyphosate toxicity. *Frontiers in Public Health, 5*, 7.

Mesnage, R., & Antoniou, M. N. (2018). Ignoring adjuvant toxicity falsifies the safety profile of commercial pesticides. *Frontiers in Public Health, 5*, 8.

Moosa, A., Shu, H., Sarachana, T., & Hu, V. W. (2018). Are endocrine disrupting compounds environmental risk factors for autism spectrum disorder? *Hormones and Behavior, 101*, 13–21.

Moser, V., Morris-Schafer, K., Richardson, J., & Li, A. (2022a). Expert panel review of the neurotoxicology of Glyphosate in mammalian systems for risk assessment purposes. *Toxicology Letters, 368*, S80–S81.

Moser, V. C., Morris-Schaffer, K., Richardson, J. R., & Li, A. A. (2022b). Glyphosate and neurological outcomes: A systematic literature review of animal studies. *Journal of Toxicology and Environmental Health-Part B-Critical Reviews, 25*(4), 162–209.

Ojiro, R., Okano, H., Takahashi, Y., Takashima, K., Tang, Q., Ozawa, S., Zou, X. Y., Woo, G. H., & Shibutani, M. (2023). Comparison of the effect of glyphosate and glyphosate-based herbicide on hippocampal neurogenesis after developmental exposure in rats. *Toxicology, 483*, 17.

Perrin, L., Spinosi, J., Chaperon, L., Kab, S., Moisan, F., & Elbaz, A. (2019). Expenditures on pesticides and incidence of Parkinson's disease in French farmers: A nationwide study. *Movement Disorders, 34*, S755–S756.

Perrin, L., Spinosi, J., Chaperon, L., Jezewski-Serra, D., El Yamani, M., Moisan, F., & Elbaz, A. (2020). Association of environmental exposure to herbicides, fungicides, and insecticides used in agriculture with incidence of Parkinson's disease in the French general population. *Movement Disorders, 35*, S82–S82.

Richardson, J. R., Fitsanakis, V., Westerink, R. H. S., & Kanthasamy, A. G. (2019). Neurotoxicity of pesticides. *Acta Neuropathologica, 138*(3), 343–362.

Samsel, A., & Seneff, S. (2013). Glyphosate's suppression of cytochrome P450 enzymes and amino acid biosynthesis by the gut microbiome: Pathways to modern diseases. *Entropy, 15*(4), 1416–1463.

Sealey, L. A., Hughes, B. W., Sriskanda, A. N., Guest, J. R., Gibson, A. D., Johnson-Williams, L., Pace, D. G., & Bagasra, O. (2016). Environmental factors in the development of autism spectrum disorders. *Environment International, 88*, 288–298.

von Ehrenstein, O. S., Ling, C. X., Cui, X., Cockburn, M., Park, A. S., Yu, F., Wu, J., & Ritz, B. (2019). Prenatal and infant exposure to ambient pesticides and autism spectrum disorder in children: Population based case-control study. *BMJ-British Medical Journal, 364*, 13.

Wanneveich, M., Moisan, F., Jacqmin-Gadda, H., Elbaz, A., & Joly, P. (2018). Projections of prevalence, lifetime risk, and life expectancy of Parkinson's disease (2010–2030) in France. *Movement Disorders, 33*(9), 1449–1455.

Winstone, J. K., Pathak, K. V., Winslow, W., Piras, I. S., White, J., Sharma, R., Huentelman, M. J., Pirrotte, P., & Velazquez, R. (2024). Glyphosate infiltrates the brain and increases proinflammatory cytokine TNFα: Implications for neurodegenerative disorders (vol 21, 20 2024). *Journal of Neuroinflammation, 21*(1), 3.

Zhou, W., Li, M. M., & Achal, V. (2025). A comprehensive review on environmental and human health impacts of chemical pesticide usage. *Emerging Contaminants, 11*(1), 13.

Chapter 6
Glyphosate and the Microbiome

The human microbiome is the subject of intense ongoing scientific research, and its central role in bodily function has gradually been discovered. This chapter describes how pesticides in general and glyphosate in particular interact with the microbiome.

Basics of the Microbiome

The *human microbiome* refers to the microorganisms—bacteria, viruses, fungi, archaea, and protozoa—that reside in various parts of the body, such as the gut, skin, mouth, respiratory tract, and urogenital tract. It is a very delicate, well-balanced system of multiple species, and its importance has become clear in recent decades (Athanasopoulou et al., 2023; Huttenhower et al., 2012; Ursell et al., 2012, 2014). In what follows, I will focus on the gut microbiome with special attention to the bacteria. Most of the gut bacteria are human-friendly, i.e., they do not cause diseases in healthy individuals; the name for this group is commensals. Some others, however, can be harmful to humans, but their activity is controlled by the whole microbiome. Notably, this is the case for healthy people; in vulnerable people (e.g., after chemotherapy or reduced immunocompetence), harmless bacteria might cause disease. This aspect is not discussed.

Disruption of the balance of the microbiome, known as dysbiosis, can be triggered by a large number of extracorporeal factors, such as antibiotic use, poor diet, infections, stress, and environmental chemicals. Dysbiosis is linked to diseases such as inflammatory bowel disease (IBD), obesity, allergies, diabetes, and even mental health disorders. While dysbiosis is associated with chronic diseases, it is also linked to acute infections, such as bacterial vaginosis or the overgrowth of the population with *Clostridium difficile* and the development of colon cancer.

The most well-known function of the gut microbiome is digestion. Gut bacteria help the breakdown of fats, complex carbohydrates, and fibers that human enzymes

N. Fraeyman, *Glyphosate 2023–2033*,
https://doi.org/10.1007/978-3-031-97540-0_6

cannot process, and they stimulate the synthesis of essential vitamins (such as vitamin K and certain B vitamins). They contribute to the metabolism of various compounds, and fermentation produces short-chain fatty acids (SCFAs), such as butyrate, which nourish colon cells and maintain gut integrity. The microbiome helps to train immune cells to distinguish between harmful pathogens and harmless substances, ensuring that the immune system responds appropriately to threats and avoids inflammation that can lead to autoimmune conditions or allergies. Each individual has a unique age-dependent composition of microorganisms that is influenced by genetic, environmental, and lifestyle factors such as diet, antibiotic use, and geography. The gut microbiome is the largest and most diverse microbiome in the human body and contains trillions of microorganisms. The role of the microbiome for human health and diseases is summarised in Fig. 6.1.

It is becoming increasingly clear that the gut microbiome balance is crucial for overall health. Dysbiosis can lead to a wide array of diseases, making the microbiome a critical area of study for improving human health and developing new therapeutic approaches. A recent area of scientific research has focused on the link between the microbiome and the brain, known as the gut–brain axis. Microbial signals can influence mood, behavior, and cognition, which are typically central nervous phenomena, but these signals are also linked to neurological diseases such as depression, anxiety, and Parkinson's disease. A comprehensive introduction to the

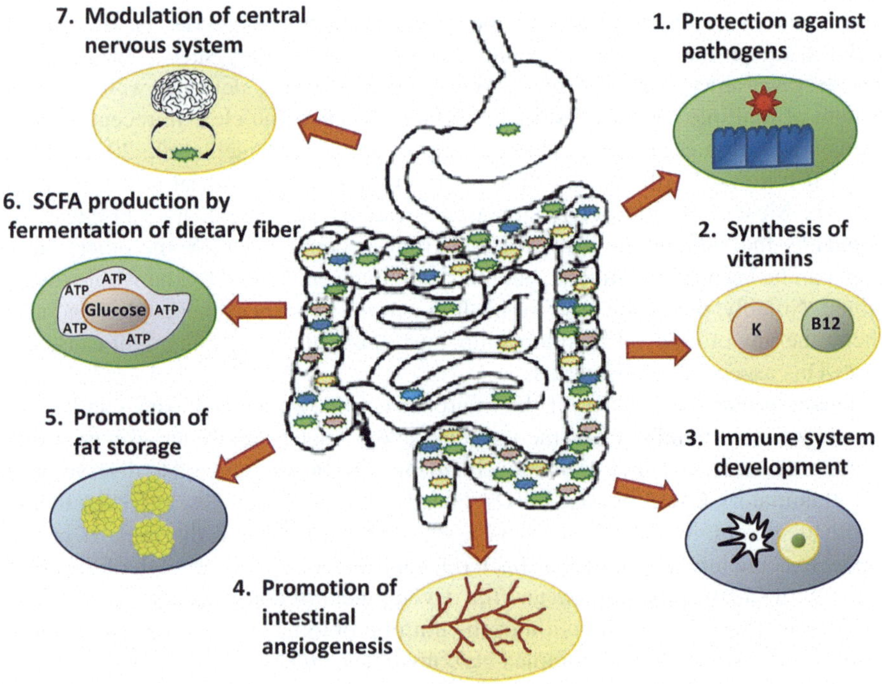

Fig. 6.1 Schematic presentation of the intestinal system and the possible influence on other bodily functions

microbiome is available: https://bio.libretexts.org/Bookshelves/Microbiology/Microbiomes_-_Health_and_the_Environment_(Parks)/01%3A_An_Introduction_to_Microbiomes/1.01%3A_An_Introduction_to_Microbiomes

The Microbiome Is Affected by Glyphosate: Neurological Implications

The next question is how the microbiome balance might be disturbed, and this is where pesticides and glyphosate come into the picture (Barnett et al., 2022; Giambo et al., 2022; Ruuskanen et al., 2023). Glyphosate was shown to be toxic to some commensal bacteria, whereas *Clostridium difficile* was resistant. This resulted in a disbalance with *C. difficile* in higher proportions, and the production of toxins (Toxin A or TcdA and toxin B or TcdB) possibly affecting the central nervous system (Hu et al., 2021; Madani & Carpenter, 2022; Mathew et al., 2022; Mesnage & Antoniou, 2020; Mesnage et al., 2022; Puigbò et al., 2022). According to Barnett and Gibson (2020), glyphosate affects EPSPS-enzyme type I, which is sensitive to glyphosate and is present in commensals, whereas type II EPSPS, which is present in some opportunistic microbes, is resistant to glyphosate. The abundant *E. coli* bacteria, which are harmless under normal conditions, can acquire mutations that induce resistance to glyphosate, hence leading to overgrowth of the bacteria and dysbiosis (Bote et al., 2019). Notably, no dysbiosis is observed in cows after exposure to glyphosate (Billenkamp et al., 2021).

There is a link between the microbiome and the central nervous system (Athanasopoulou et al., 2023; Huttenhower et al., 2012; Lin et al., 2023; Lynch & Pedersen, 2016). Dysbiosis has been shown to cause central neurological and peripheral diseases (Cryan et al., 2020; Kho & Lal, 2018; Lynch & Pedersen, 2016; Sorboni et al., 2022) such as depression (Bastiaanssen et al., 2020), deficiencies of the immune system (Huang et al., 2020), allergies (Lynch, 2016), Parkinson's disease (Tan et al., 2021), and autism spectrum disorders (Chernikova et al., 2021; Dovrolis et al., 2019; Saurman et al., 2020; Taniya et al., 2022). The link between glyphosate exposure and delayed neurodevelopment in children is suggested (Etiemble & Cordier, 2022). A recent overview presented the links between glyphosate, the gut microbiome, and the health of experimental animals (mice), earthworms, and honey bees (Walsh et al., 2023).

Glyphosate can enter the central nervous system by crossing the blood–brain barrier—see Chap. 3 (Winstone et al., 2022)—and possibly affect functionality directly, whereas dysbiosis has an indirect effect. Alternatively, a direct effect on the gut microbiome and the consequences of the immune-endocrine system have been described (Madani & Carpenter, 2022). Another hypothesis is that dysbiosis leads to endotoxemia, i.e., the leakage of endotoxins such as lipopolysaccharides from the gut, which in turn activate inflammatory cascades, eventually in the brain (Izumi et al., 2023, 2024).

The study of the impact of glyphosate on the gut microbiome has taken one major step forward through the application of modern genome technology, including shotgun metagenomics (Lehman et al., 2023; Mesnage et al., 2021; Simoes et al., 2018). These results demonstrate the selective effect of glyphosate on bacterial species within the gut of rats after 90 days of treatment.

In 2013, the scientific world was baffled by a paper by Samsel and Seneff (Samsel & Seneff, 2013), who claimed that glyphosate was the cause of autism, and the authors predicted that by 2030, 50% of children would be autistic; this is discussed in Chap. 5. Although the paper was judged as scientifically incorrect shortly after publication, a sense of discomfort remained, and further studies were requested. Experiments in rats demonstrated a possible link between exposure to glyphosate and the rat-equivalent phenotype of human autism spectrum disease (ASD), and soluble epoxide hydrolase was suggested as the biological target (Pu et al., 2021; Pu et al., 2020). Overt ASD symptoms in humans have been described (Ongono et al., 2020; von Ehrenstein et al., 2019), although glyphosate is considered not only the cause but also other chemicals, pesticides, fragrances, and other adjuvants.

As usual, scientific research is subject to criticism, which forces the search for better postulates and answers to (medical) questions. The existence of the gut–brain axis is beyond doubt, and disturbing the balance between commensal and obligatory pathogens has far-reaching consequences for human health. The link between the microbiome and diseases and the role of, e.g., glyphosate or other pesticides in the human enteric system are more questioned (Barnett & Gibson, 2020). The role of pesticides is probably deterministic, although pesticides are likely to be part of the total exposure to chemicals. Hence, the blame is on the exposome (see Chap. 7) in which glyphosate is one component. The abundant use of this herbicide worldwide has led to its presence in all compartments of the environment, making it one of the principal chemicals we are all exposed to.

The European Food Safety Authority is concerned about the impact that exposure to chemicals via food, pesticides, and other sources might have on the microbiome; two recent documents are exemplary.

In 2020, an editorial entitled "Exploring the need to include microbiomes into EFSA's scientific assessments" was published (https://efsa.onlinelibrary.wiley.com/doi/10.2903/j.efsa.2020.e18061). The main ideas are that (1) the microbiome is individual and hence highly diverse, (2) there is yet no standard definition of a healthy microbiome, and (3) it is not possible to distinguish direct from indirect effects of extracorporeal compounds on the microbiome. The EFSA doubts whether they should define "a healthy microbiome." They refer to recent developments, including the use of advanced gen-technology to unravel the role of the microbiome.

The second document was published in 2024: "Roadmap for the integration of gastrointestinal (GI) tract microbiomes (human and domestic animals) in risk assessments under EFSA's remit" (https://www.efsa.europa.eu/en/supporting/pub/en-8597). In this 238-page long document, the lack of basic knowledge on the role of the microbiome in health and disease is repeated owing to the inconsistency in methodologies, lack of clinical data, and inconsistencies of the experimental results. According to the authors, the focus should be on three fundamental aspects: (1) the need to understand the fundamental molecular mechanisms, (2) the need for harmonization and standardization to study the microbiome, and (3) the extrapolation of the results of in vitro and in vivo animal studies to the human microbiome.

The interest of the EFSA to include the microbiome in the evaluation process of chemicals is appreciated. However, in both documents, the lack of sufficient knowledge of the molecular mechanisms and the lack of universally accepted methods and standards for the study is considered sufficient to conclude that microbiome studies should NOT be included in the glyphosate assessment process. Yet, the actual evidence for the involvement of the microbiome in health and disease is overwhelming, although the underlying molecular mechanisms are not fully elucidated. The question remains as to how much evidence is needed, and because scientific research cannot be forced to deliver on time, it could take a long time before evidence is gathered. Given the uncertainty, this is a situation where the precautionary principle should fully be applicable. The EFSA is well aware of this problem.

Taken together, a direct link between glyphosate and ASD in humans seems unlikely, but scientific evidence is accumulating, and further basic and epidemiological research is needed. Although the claim of Samsel and Seneff was not justified, the possibility that exogenous chemicals such as pesticides in the exposome are linked to neurological disorders should not be ignored.

References

Athanasopoulou, K., Adamopoulos, P. G., & Scorilas, A. (2023). Unveiling the human gastrointestinal tract microbiome: The past, present, and future of metagenomics. *Biomedicine, 11*(3), 16. https://doi.org/10.3390/biomedicines11030827

Barnett, J. A., & Gibson, D. L. (2020). Separating the empirical wheat from the pseudoscientific chaff: A critical review of the literature surrounding glyphosate, dysbiosis and wheat-sensitivity. *Frontiers in Microbiology, 11*, 8. https://doi.org/10.3389/fmicb.2020.556729

Barnett, J. A., Bandy, M. L., & Gibson, D. L. (2022). Is the use of glyphosate in modern agriculture resulting in increased neuropsychiatric conditions through modulation of the gut-brain-microbiome axis? *Frontiers in Nutrition, 9*, 9. https://doi.org/10.3389/fnut.2022.827384

Bastiaanssen, T. F. S., Cussotto, S., Claesson, M. J., Clarke, G., Dinan, T. G., & Cryan, J. F. (2020). Gutted! Unraveling the role of the microbiome in major depressive disorder. *Harvard Review of Psychiatry, 28*(1), 26–39. https://doi.org/10.1097/hrp.0000000000000243

Billenkamp, F., Schnabel, K., Huther, L., Frahm, J., von Soosten, D., Meyer, U., et al. (2021). No hints at glyphosate-induced ruminal dysbiosis in cows. *npj Biofilms and Microbiomes, 7*(1), 13. https://doi.org/10.1038/s41522-021-00198-4

Bote, K., Pöppe, J., Merle, R., Makarova, O., & Roesler, U. (2019). Minimum inhibitory concentration of glyphosate and of a glyphosate-containing herbicide formulation for *Escherichia coli* isolates-differences between pathogenic and non-pathogenic isolates and between host species. *Frontiers in Microbiology, 10*, 11. https://doi.org/10.3389/fmicb.2019.00932

Chernikova, M. A., Flores, G. D., Kilroy, E., Labus, J. S., Mayer, E. A., & Aziz-Zadeh, L. (2021). The brain-gut-microbiome system: Pathways and implications for autism spectrum disorder. *Nutrients, 13*(12), 23. https://doi.org/10.3390/nu13124497

Cryan, J. F., O'Riordan, K. J., Sandhu, K., Peterson, V., & Dinan, T. G. (2020). The gut microbiome in neurological disorders. *Lancet Neurology, 19*(2), 179–194. https://doi.org/10.1016/s1474-4422(19)30356-4

Dovrolis, N., Kolios, G., Spyrou, G. M., & Maroulakou, I. (2019). Computational profiling of the gut-brain axis: Microflora dysbiosis insights to neurological disorders. *Briefings in Bioinformatics, 20*(3), 825–841. https://doi.org/10.1093/bib/bbx154

Etiemble, J., & Cordier, S. (2022). Pesticides and neurodevelopment in children. *Environnement Risques & Sante, 21*(1), 51–65. https://doi.org/10.1684/ers.2021.1615

Giambo, F., Costa, C., Teodoro, M., & Fenga, C. (2022). Role-playing between environmental pollutants and human gut microbiota: A complex bidirectional interaction. *Frontiers in Medicine, 9*, 10. https://doi.org/10.3389/fmed.2022.810397

Hu, J. Z., Lesseur, C., Miao, Y., Manservisi, F., Panzacchi, S., Mandrioli, D., et al. (2021). Low-dose exposure of glyphosate-based herbicides disrupt the urine metabolome and its interaction with gut microbiota. *Scientific Reports, 11*(1), 10. https://doi.org/10.1038/s41598-021-82552-2

Huang, Q., Yu, F., Liao, D., & Xia, J. (2020). Microbiota-immune system interactions in human neurological disorders. *CNS & Neurological Disorders-Drug Targets, 19*(7), 509–526. https://doi.org/10.2174/1871527319666200726222138

Huttenhower, C., Gevers, D., Knight, R., Abubucker, S., Badger, J. H., Chinwalla, A. T., et al. (2012). Structure, function and diversity of the healthy human microbiome. *Nature, 486*(7402), 207–214. https://doi.org/10.1038/nature11234

Izumi, Y., O'Dell, K. A., & Zorumski, C. F. (2023). The herbicide glyphosate inhibits hippocampal long-term potentiation and learning through activation of pro-inflammatory signaling. *Scientific Reports, 13*(1), 12. https://doi.org/10.1038/s41598-023-44121-7

Izumi, Y., O'Dell, K. A., & Zorumski, C. F. (2024). Glyphosate as a direct or indirect activator of pro-inflammatory signaling and cognitive impairment. *Neural Regeneration Research, 19*(10), 2212–2218. https://doi.org/10.4103/1673-5374.391331

Kho, Z. Y., & Lal, S. K. (2018). The human gut microbiome – a potential controller of wellness and disease. *Frontiers in Microbiology, 9*, 23. https://doi.org/10.3389/fmicb.2018.01835

Lehman, P. C., Cady, N., Ghimire, S., Shahi, S. K., Shrode, R. L., Lehmler, H. J., & Mangalam, A. K. (2023). Low-dose glyphosate exposure alters gut microbiota composition and modulates gut homeostasis. *Environmental Toxicology and Pharmacology, 100*, 9. https://doi.org/10.1016/j.etap.2023.104149

Lin, M. S., Wang, Y. C., Chen, W. J., & Kung, W. M. (2023). Impact of gut-brain interaction in emerging neurological disorders. *World Journal of Clinical Cases, 11*(1), 1–6. https://doi.org/10.12998/wjcc.v11.i1.1

Lynch, S. V. (2016). Gut microbiota and allergic disease new insights. *Annals of the American Thoracic Society, 13*, S51–S54. https://doi.org/10.1513/AnnalsATS.201507-451MG

Lynch, S. V., & Pedersen, O. (2016). The human intestinal microbiome in health and disease. *New England Journal of Medicine, 375*(24), 2369–2379. https://doi.org/10.1056/NEJMra1600266

Madani, N. A., & Carpenter, D. O. (2022). Effects of glyphosate and glyphosate-based herbicides like RoundupTM on the mammalian nervous system: A review. *Environmental Research, 214*, 9. https://doi.org/10.1016/j.envres.2022.113933

Mathew, S. A., Muola, A., Saikkonen, K., Saloniemi, I., Helander, M., & Puigbo, P. (2022). Quantification of the potential impact of glyphosate-based products on microbiomes. *JoVE-Journal of Visualized Experiments, 179*, 26. https://doi.org/10.3791/63109

Mesnage, R., & Antoniou, M. N. (2020). Computational modelling provides insight into the effects of glyphosate on the shikimate pathway in the human gut microbiome. *Current Research in Toxicology, 1*, 25–33. https://doi.org/10.1016/j.crtox.2020.04.001

Mesnage, R., Teixeira, M., Mandrioli, D., Falcioni, L., Ducarmon, Q. R., Zwittink, R. D., et al. (2021). Use of shotgun metagenomics and metabolomics to evaluate the impact of glyphosate or roundup MON 52276 on the gut microbiota and serum metabolome of Sprague-Dawley rats. *Environmental Health Perspectives, 129*(1), 15. https://doi.org/10.1289/ehp6990

Mesnage, R., Bowyer, R. C. E., El Balkhi, S., Saint-Marcoux, F., Gardere, A., Ducarmon, Q. R., et al. (2022). Impacts of dietary exposure to pesticides on faecal microbiome metabolism in adult twins. *Environmental Health, 21*(1), 46. https://doi.org/10.1186/s12940-022-00860-0

Ongono, J. S., Béranger, R., Baghdadli, A., & Mortamais, M. (2020). Pesticides used in Europe and autism spectrum disorder risk: Can novel exposure hypotheses be formulated beyond organophosphates, organochlorines, pyrethroids and carbamates? – a systematic review. *Environmental Research, 187*, 20. https://doi.org/10.1016/j.envres.2020.109646

Pu, Y. Y., Yang, J., Chang, L. J., Qu, Y. G., Wang, S. M., Zhang, K., et al. (2020). Maternal glyphosate exposure causes autism-like behaviors in offspring through increased expression of soluble epoxide hydrolase. *Proceedings of the National Academy of Sciences of the United States of America, 117*(21), 11753–11759. https://doi.org/10.1073/pnas.1922287117

Pu, Y. Y., Ma, L., Shan, J. J., Wan, X. Y., Hammock, B. D., & Hashimoto, K. (2021). Autism-like behaviors in male juvenile offspring after maternal glyphosate exposure. *Clinical Psychopharmacology and Neuroscience, 19*(3), 554–558. https://doi.org/10.9758/cpn.2021.19.3.554

Puigbò, P., Leino, L. I., Rainio, M. J., Saikkonen, K., Saloniemi, I., & Helander, M. (2022). Does glyphosate affect the human microbiota? *Life-Basel, 12*(5), 8. https://doi.org/10.3390/life12050707

Ruuskanen, S., Fuchs, B., Nissinen, R., Puigbo, P., Rainio, M., Saikkonen, K., & Helander, M. (2023). Ecosystem consequences of herbicides: The role of microbiome. *Trends in Ecology & Evolution, 38*(1), 35–43. https://doi.org/10.1016/j.tree.2022.09.009

Samsel, A., & Seneff, S. (2013). Glyphosate's suppression of cytochrome P450 enzymes and amino acid biosynthesis by the gut microbiome: Pathways to modern diseases. *Entropy, 15*(4), 1416–1463. https://doi.org/10.3390/e15041416

Saurman, V., Margolis, K. G., & Luna, R. A. (2020). Autism spectrum disorder as a brain-gut-microbiome axis disorder. *Digestive Diseases and Sciences, 65*(3), 818–828. https://doi.org/10.1007/s10620-020-06133-5

Simoes, T., Novais, S. C., Natal-Da-Luz, T., Devreese, B., de Boer, T., Roelofs, D., et al. (2018). An integrative omics approach to unravel toxicity mechanisms of environmental chemicals: Effects of a formulated herbicide. *Scientific Reports, 8*, 12. https://doi.org/10.1038/s41598-018-29662-6

Sorboni, S. G., Moghaddam, H. S., Jafarzadeh-Esfehani, R., & Soleimanpour, S. (2022). A comprehensive review on the role of the gut microbiome in human neurological disorders. *Clinical Microbiology Reviews, 35*(1), 50. https://doi.org/10.1128/cmr.00338-20

Tan, A. H., Chong, C. W., Lim, S. Y., Yap, I. K. S., Teh, C. S. J., Loke, M. F., et al. (2021). Gut microbial ecosystem in Parkinson disease: New clinicobiological insights from multi-omics. *Annals of Neurology, 89*(3), 546–559. https://doi.org/10.1002/ana.25982

Taniya, M. A., Chung, H. J., Al Mamun, A., Alam, S., Aziz, M. A., Emon, N. U., et al. (2022). Role of gut microbiome in autism spectrum disorder and its therapeutic regulation. *Frontiers in Cellular and Infection Microbiology, 12*, 13. https://doi.org/10.3389/fcimb.2022.915701

Ursell, L. K., Metcalf, J. L., Parfrey, L. W., & Knight, R. (2012). Defining the human microbiome. *Nutrition Reviews, 70*, S38–S44. https://doi.org/10.1111/j.1753-4887.2012.00493.x

Ursell, L. K., Haiser, H. J., Van Treuren, W., Garg, N., Reddivari, L., Vanamala, J., et al. (2014). The intestinal metabolome: An intersection between microbiota and host. *Gastroenterology, 146*(6), 1470–1476. https://doi.org/10.1053/j.gastro.2014.03.001

von Ehrenstein, O. S., Ling, C. X., Cui, X., Cockburn, M., Park, A. S., Yu, F., et al. (2019). Prenatal and infant exposure to ambient pesticides and autism spectrum disorder in children: Population based case-control study. *BMJ-British Medical Journal, 364*, 13. https://doi.org/10.1136/bmj.l962

Walsh, L., Hill, C., & Ross, R. P. (2023). Impact of glyphosate (Roundup™) on the composition and functionality of the gut microbiome. *Gut Microbes, 15*(2), 16. https://doi.org/10.1080/19490976.2023.2263935

Winstone, J. K., Pathak, K. V., Winslow, W., Piras, I. S., White, J., Sharma, R., et al. (2022). Glyphosate infiltrates the brain and increases pro-inflammatory cytokine TNF alpha: Implications for neurodegenerative disorders. *Journal of Neuroinflammation, 19*(1), 14. https://doi.org/10.1186/s12974-022-02544-5

Chapter 7
Broaden the View on Glyphosate, the Exposome

Glyphosate, like many other chemicals, has become part of the environment. Thus, it is imperative to study glyphosate as one component of this mixture, and the properties of the mixture are not equal to the sum of the individual components. In this chapter, some details on the mixture idea are given together with the role of glyphosate. In addition, some ways to cope with this problem are summarized.

Basics on Mixtures of Chemicals

The whole world is exposed to a myriad of chemicals denoted as "the chemical anthropocene" (Arcuri & Hendlin, 2019). Most of these chemicals have unknown (eco)toxicological profiles. A recent inventory of CAS-numbered chemicals revealed 157,000 identified and 50,000–70,000 unidentified chemicals (Wang et al., 2020). Whatever the exact number is, bodily exposure to chemicals is abundant and complex. Notably, the environmental concentrations are usually very low, mostly at the limits of analytical/chemical detection methods.

The problem of mixtures of pollutants was conceptualized in 2005 by Wild, who defined the "exposome" as the sum of environmental extra-corporal chemicals and the endogenous molecules in the body reacting upon exposure to these external pollutants, excluding genomic variables but including socioeconomic variables (Vermeulen et al., 2020; Wild, 2012). These chemicals act together mostly in an unknown way, contradicting the assumption that the total effect is the simple sum of all the individual effects. At least three different types of interactions between the pollutants are distinguished. (i) Physically, the combination of lipophilic (e.g., dioxins) and hydrophilic pollutants (e.g., glyphosate) could enhance the toxic effect of the hydrophilic substances via the membranous effect of the lipophilic substances (e.g., tallowamine); (ii) toxic-pharmacokinetic, i.e., inhibition or stimulation of the metabolizing enzymes in the liver (the CYT P450 system) could decrease or increase

N. Fraeyman, *Glyphosate 2023–2033*,
https://doi.org/10.1007/978-3-031-97540-0_7

the activity of any individual pollutant, drug or endogenous molecule; and (iii) pharmacodynamic, i.e., interactions at the cellular receptors and signal transduction systems might induce unexpected tissue reactivity, e.g., glyphosate may interact with protein kinase C (Cavalli et al., 2013). Some of the possibilities have been demonstrated in experimental conditions; it remains to be proven that similar mechanisms are or could be operational in real-life conditions (Camele et al., 2022). Evidence for complex interactions was obtained by Docea et al. (Docea et al., 2018), who investigated the effects of a 6-month exposure to 13 chemicals at the NOAEL level. These authors concluded that the mutual modulation of different mechanisms could explain the results. According to some, the exposome has the same complexity as the genome (Vermeulen et al., 2020). Certainly, human health is threatened not by one chemical but by many simultaneously. The importance for human health was acknowledged about 10 years ago (Svingen & Vinggaard, 2016), and the link between diseases in children and exposure to manufactured chemicals requires, according to the Consortium for Children's Environmental Health, new legislation as published in the *New England Journal of Medicine* (https://www.nejm.org/doi/full/10.1056/NEJMms2409092).

The difficulty of studying the ultimate effects of mixtures of chemicals was already acknowledged by the European Scientific Committee on Health and Environmental Risks (SCHER) in 2012 (https://data.europa.eu/doi/10.2772/21444). More recently, the European Food Safety Authority (EFSA) developed a methodology known as the Monte Carlo Risk Assessment (MCRA) web platform (https://mcra.rivm.nl/mcra/#/) to tackle the problem of the effect of mixtures of chemicals called the "cocktail effects," which is approached through a cumulative risk assessment (CRA). The MRCA platform presents a user guide for performing research in the field, organized at the European level, and contains data from recent scientific research, examples, calculation tools, etc. Figure 7.1 shows the many "modules" within the MCRA program, illustrating the complexity of the problem and the complexity of the proposed handling method. The ongoing project is PARC, running from 2022 until 2029 (https://www.eu-parc.eu/).

EFSA published in 2019 a scientific report entitled: "Establishment of cumulative assessment groups of pesticides for their effects on the thyroid." In this 50-page document, the effects of 128 active substances on hypothyroidism and 17 on C-cell hypertrophy, hyperplasia, and neoplasia were summarized together with the NOAELs. In the summary, it is stated: "*Based on the number and NOAEL's of the active substances in each of the cumulative assessment groups, performing a cumulative risk assessment is expected to cover the combined effects of pesticides*." The authors avoid using concepts such as synergism, interaction, and antagonism and assume that the final effect is the sum of the individual effects. Some very recent work stresses the need to consider the concept of exposome when studying thyroid function and, by extrapolation, when considering the toxic effect on the endocrine system as a whole (Lee et al., 2025).

One particular and interesting item in this EFSA text is point 3.1, the precautionary principle and uncertainties. (p5 in the document): *Regulation (EC) No 1107/2009 concerning the placing of plant protection products on the market does not*

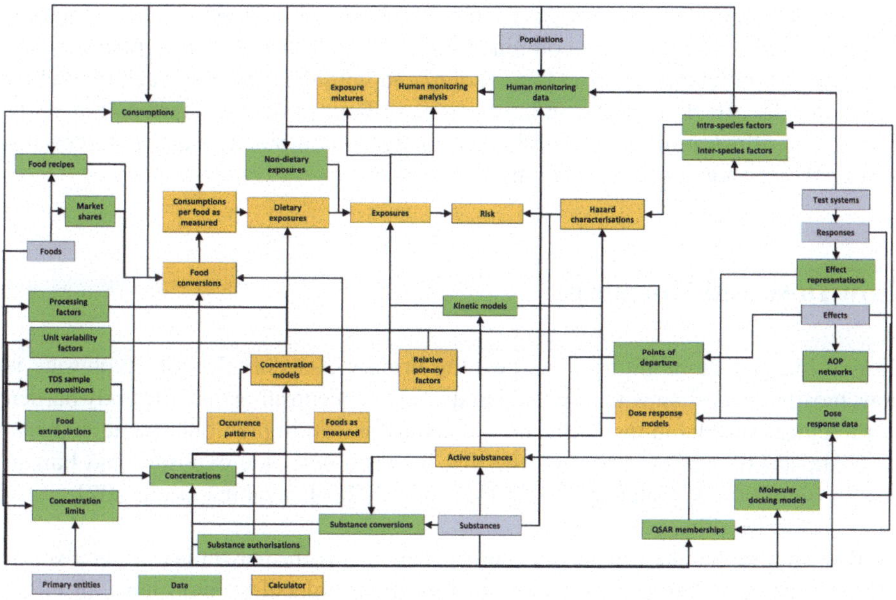

Fig. 7.1 Modules and their interaction within the MCRA program

prescribe precisely how CRA of pesticides should be performed. However, it provides that Member States "shall not be prevented from applying the precautionary principle where there is scientific uncertainty as to the risks with regard to human and animal health" and "shall take into consideration possible elements of uncertainty in the information in order to ensure that the chances of failing to detect adverse effects or of underestimating their importance are reduced to a minimum." This illustrates the awareness of EFSA for the precautionary principle approach but—at the same time—the responsibility of applying the principle is shifted to the member states and the legal background (regulation 1107/2009) is used as an umbrella-type shield for not having to make any decision. Given the scientific reputation of EFSA/ECHA, a more convincing attitude is expected.

Glyphosate and Exposome

Together with other chemicals, such as BPA, DDT (which is still present in the environment), and members of the PFAS family (because of their resistance to degradation), glyphosate is likely to be a prominent member of the exposome because of its widespread use and omnipresence in every compartment of the environment (Samanipour et al., 2024). How glyphosate interacts with other chemicals is essentially unknown, although, on the basis of its biological effects, as evidenced under experimental conditions, interactions with other chemicals, e.g., with endogenous agonists at steroid receptors or in the synthesis of estrogens, are expected. Experimental data indicate a combined effect of endotoxins and glyphosate, two common air pollutants in agriculture. Compared with the effects of each individual pollutant, the effects (cellular inflammation) of both pollutants were significantly synergistic, illustrating the possibility of unwanted combined effects (Merkowsky et al., 2020; Pandher et al., 2021a, b).

Low Dose and Hormesis

In real-life and biomonitoring studies, the urinary concentration of glyphosate is low, mostly lower than what is considered a safe concentration, and the body burden is therefore considered insignificant. As already shown in previous chapters, this is in contrast with recent papers on the link between pesticide exposure and human health (Guillotin & Delcourt, 2022; He et al., 2022; Kalyabina et al., 2021; Kaur et al., 2024; Mattila et al., 2021).

Research in the late nineteenth century, which was further refined in the twentieth century, revealed that low doses of bioactive molecules could have an effect opposite to the toxic effect that occurs at higher concentrations/doses; this effect is referred to as "hormesis." The concept was rediscovered in modern pharmacology (Calabrese & Baldwin, 2003) and later for pesticides (Calabrese, 2015; Duke, 2014) and agriculture (Jalal et al., 2021; Xu et al., 2023). Although hormesis is fairly well described in some experimental models, the existence of a hormetic effect is questioned by others because the effect seems less reproducible than the effect at high doses, i.e., the full toxic effect. Furthermore, there is no readily acceptable explanation for this low-dose effect. Nevertheless, despite these restrictions, hormesis should be taken into account.

Taken together, exposure to many compounds, including pesticides, albeit at low doses, should be considered harmful, although the inherent complexity of this situation hinders the elucidation of the mechanisms of the interactions leading to a definite phenotype. In such cases, the precautionary principle should be applied. Major efforts are ongoing to elucidate the identification of chemicals or groups of chemicals in complex mixtures (Escher et al., 2020; Svingen & Vinggaard, 2016). A pragmatic approach promoted by Boberg et al. is based on reducing the complexity of the exposome by grouping pollutants with similar properties using QSAR (quantitative structure–activity relationship)-based methodology (Boberg et al., 2019). Considering the multitude of pollutants will also necessitate the recalculation of variables, such as ADI and NOAEL. This has been studied recently (Agathokleous et al., 2021).

From a philosophical and ethical point of view, it is important to avoid additional molecules from those already present, and it is indicated to remove whenever possible, molecules from the mixture. The reasoning that since there is already so much on the market, one more will not make a difference is at least worrying, rather unethical; some call it criminal. It fuels the impression that there is little to be done except to note that health is possibly jeopardized. Ironically, no one, irrespective of socioeconomic status, including those who manufacture the chemicals and have the largest economic benefit, escapes exposure.

References

Agathokleous, E., Moore, M. N., & Calabrese, E. J. (2021). Estimating the no-observed-adverse-effect-level (NOAEL) of hormetic dose-response relationships in meta-data evaluations. *MethodsX, 8*. https://doi.org/10.1016/j.mex.2021.101568

Arcuri, A., & Hendlin, Y. H. (2019). The chemical Anthropocene: Glyphosate as a case study of pesticide exposures. *King's Law Journal, 30*(2), 234–253. https://doi.org/10.1080/09615768.2019.1645436

Boberg, J., Dybdahl, M., Petersen, A., Hass, U., Svingen, T., & Vinggaard, A. M. (2019). A pragmatic approach for human risk assessment of chemical mixtures. *Current Opinion in Toxicology, 15*, 1–7. https://doi.org/10.1016/j.cotox.2018.11.004

Calabrese, E. J. (2015, August 16–20). *Perspectives on hormesis and implications for pesticides.* Paper presented at the symposium on pesticide dose: Effects on the environment and target and non-target organisms/250th ACS national meeting and exposition, Boston, MA.

Calabrese, E. J., & Baldwin, L. A. (2003). Chemotherapeutics and hormesis. *Critical Reviews in Toxicology, 33*(3–4), 305–353. https://doi.org/10.1080/713611041

Camele, G., Menazzi, S., Chanfreau, H., Marraco, A., Hasperue, W., Butti, M. D., & Abba, M. C. (2022). Multiomix: A cloud-based platform to infer cancer genomic and epigenomic events associated with gene expression modulation. *Bioinformatics, 38*(3), 866–868. https://doi.org/10.1093/bioinformatics/btab678

Cavalli, V., Cattani, D., Rieg, C. E. H., Pierozan, P., Zanatta, L., Parisotto, E. B., et al. (2013). Roundup disrupts male reproductive functions by triggering calcium-mediated cell death in rat testis and Sertoli cells. *Free Radical Biology and Medicine, 65*, 335–346. https://doi.org/10.1016/j.freeradbiomed.2013.06.043

Docea, A. O., Gofita, E., Goumenou, M., Calina, D., Rogoveanu, O., Varut, M., et al. (2018). Six months exposure to a real life mixture of 13 chemicals' below individual NOAELs induced non monotonic sex-dependent biochemical and redox status changes in rats. *Food and Chemical Toxicology, 115*, 470–481. https://doi.org/10.1016/j.fct.2018.03.052

Duke, S. O. (2014). Hormesis with pesticides. *Pest Management Science, 70*(5), 689–689. https://doi.org/10.1002/ps.3756

Escher, B. I., Stapleton, H. M., & Schymanski, E. L. (2020). Tracking complex mixtures of chemicals in our changing environment. *Science, 367*(6476), 388. https://doi.org/10.1126/science.aay6636

Guillotin, S., & Delcourt, N. (2022). Studying the impact of persistent organic pollutants exposure on human health by proteomic analysis: A systematic review. *International Journal of Molecular Sciences, 23*(22), 26. https://doi.org/10.3390/ijms232214271

He, X., Tu, Y., Song, Y. W., Yang, G. H., & You, M. D. (2022). The relationship between pesticide exposure during critical neurodevelopment and autism spectrum disorder: A narrative review. *Environmental Research, 203*, 12. https://doi.org/10.1016/j.envres.2021.111902

Jalal, A., de Oliveira, J. C., Ribeiro, J. S., Fernandes, G. C., Mariano, G. G., Trindade, V. D. R., & dos Reis, A. R. (2021). Hormesis in plants: Physiological and biochemical responses. *Ecotoxicology and Environmental Safety, 207*, 12. https://doi.org/10.1016/j.ecoenv.2020.111225

Kalyabina, V. P., Esimbekova, E. N., Kopylova, K. V., & Kratasyuk, V. A. (2021). Pesticides: Formulants, distribution pathways and effects on human health-a review. *Toxicology Reports, 8*, 1179–1192. https://doi.org/10.1016/j.toxrep.2021.06.004

Kaur, R., Choudhary, D., Bali, S., Bandral, S. S., Singh, V., Ahmad, M. A., et al. (2024). Pesticides: An alarming detrimental to health and environment. *Science of the Total Environment, 915*, 17. https://doi.org/10.1016/j.scitotenv.2024.170113

Lee, J., Jang, H., Pearce, E. N., & Shin, H. M. (2025). Exposome-wide association study of thyroid function using U.S. National Health and Nutrition Examination Survey data. *Environmental Research, 269*. https://doi.org/10.1016/j.envres.2025.120884

Mattila, T., Santonen, T., Andersen, H. R., Katsonouri, A., Szigeti, T., Uhl, M., et al. (2021). Scoping review-the association between asthma and environmental chemicals. *International Journal of Environmental Research and Public Health, 18*(3), 13. https://doi.org/10.3390/ijerph18031323

Merkowsky, K., Pandher, U. S., Schneberger, D., Thompson, B., Aulakh, G. K., Singh, B., & Kirychuk, S. (2020). Synergistic lung inflammation related to agricultural co-exposures: Lipopolysaccharide (LPS) and glyphosate. *American Journal of Respiratory and Critical Care Medicine, 201*, 1. Retrieved from <Go to ISI>://WOS:000556393501196.

Pandher, U., Kirychuk, S., Schneberger, D., Thompson, B., Aulakh, G., Sethi, R. S., & Singh, B. (2021a). Lung inflammation from repeated exposure to LPS and glyphosate. *Cell and Tissue Research, 386*(3), 637–648. https://doi.org/10.1007/s00441-021-03531-7

Pandher, U., Kirychuk, S., Schneberger, D., Thompson, B., Aulakh, G., Sethi, R. S., & Singh, B. (2021b). Pulmonary inflammatory response from co-exposure to LPS and glyphosate. *Environmental Toxicology and Pharmacology, 86*, 12. https://doi.org/10.1016/j.etap.2021.103651

Samanipour, S., Barron, L. P., van Herwerden, D., Praetorius, A., Thomas, K. V., & O'Brien, J. W. (2024). Exploring the chemical space of the exposome: How far have we gone? *JACS Au, 14*. https://doi.org/10.1021/jacsau.4c00220

Svingen, T., & Vinggaard, A. M. (2016). The risk of chemical cocktail effects and how to deal with the issue. *Journal of Epidemiology and Community Health, 70*(4), 322–323. https://doi.org/10.1136/jech-2015-206268

Vermeulen, R., Schymanski, E. L., Barabási, A. L., & Miller, G. W. (2020). The exposome and health: Where chemistry meets biology. *Science, 367*(6476), 392–396. https://doi.org/10.1126/science.aay3164

Wang, Z. Y., Walker, G. W., Muir, D. C. G., & Nagatani-Yoshida, K. (2020). Toward a global understanding of chemical pollution: A first comprehensive analysis of national and regional chemical inventories. *Environmental Science & Technology, 54*(5), 2575–2584. https://doi.org/10.1021/acs.est.9b06379

Wild, C. P. (2012). The exposome: From concept to utility. *International Journal of Epidemiology, 41*(1), 24–32. https://doi.org/10.1093/ije/dyr236

Xu, N., Sun, Y., Wang, Y. R., Cui, Y. D., Jiang, Y. J., & Zhang, C. (2023). Hormesis effects in tomato plant growth and photosynthesis due to acephate exposure based on physiology and transcriptomic analysis. *Pest Management Science, 79*(6), 2029–2039. https://doi.org/10.1002/ps.7381

Chapter 8
Glyphosate and the Environment

Glyphosate has an impact on the environment, and although the focus is on human health, this effect is far from negligible. In this chapter, I first provide an overview of how to measure the impact of glyphosate on the environment; then, two aspects are discussed: the effect of glyphosate on biodiversity in general and on pollinators, particularly honey bees.

Impact of Pesticides in General and Glyphosate in Particular on Soil Organisms and How to Measure It

Knowing the extent of the (negative) impact of pesticides and glyphosate on the environment is the first step when asking for remediation. This is a difficult exercise, if possible, at all.

The first step is to analyze the presence of pesticides in the environment, i.e., in soil and water. The data are clear: Some pesticides are present in significant amounts in the environment (Badani et al., 2023; Gill et al., 2017; Klátyik et al., 2023; Mohy-Ud-Din et al., 2024). Allister and Kogan (Alister & Kogan, 2006) using an adapted mixed method, studied the physicochemical properties of 11 herbicides, 4 insecticides, and 3 fungicides, such as persistence, leaching volatility, the octanol–water coefficient, the reference dose, and the lethal dose to mammals, birds, aquatic animals, and insects, and correlated these values with experimentally obtained values for their presence in ground and surface water. They found a good correlation, allowing them to predict the environmental burden on the basis of the properties of the compounds. Extending this approach to other pesticides could help approximate possible environmental problems, eventually for new compounds.

The second point is the residence time of pesticides in the soil. This is for glyphosate highly variable between a few days and 6 months, depending on the composition of the soil and the overall atmospheric conditions. Glyphosate binds tightly to

N. Fraeyman, *Glyphosate 2023–2033*,
https://doi.org/10.1007/978-3-031-97540-0_8

clay particles, thereby losing its herbicidal properties (Hagner et al., 2019; Kanissery et al., 2019). Glyphosate is certainly not a persistent ("forever") pollutant in soils, unlike some members of the PFAS family.

The next step is to evaluate the effects on the microorganisms living in the environment.

Recent reports have shown that the impact of glyphosate is significant. A complete overview was given by Gill et al., illustrating the devastating effects on unicellular organisms (fungal species, bacteria, etc.), on multicellular organisms ranging from lower invertebrates to higher vertebrates (Gill et al., 2017) as well as on other soil organisms, such as earthworms and springtails (Altmanninger et al., 2023; Brandmaier et al., 2023). In addition, these authors reported that the formulation had a greater influence, illustrating the importance of adjuvants to glyphosate. Overall, these data indicate the vulnerability of many microorganisms to pesticides and glyphosate.

Using a complex algorithm in which data for fish, bird, bee, and arthropod toxicity were combined with data on the half-life in soil and on the plant surface, a figure for the pesticide environmental impact quotient (EIQ) was obtained (Kniss & Coburn, 2015). Multiplying the EIQ with the pesticide use rate yields a field EIQ. This field EIQ is used for comparative purposes when choosing a particular pesticide.

In addition, on the basis of 15 indices for herbicides, insecticides, and fungicides, a pesticide risk tool (PRT) was calculated (Meys et al., 2024). This tool was used for the evaluation of pesticide use in integrated pest management programs. They compared this figure with the older EIQ method and found a good overall correlation among the high, middle, and low toxicity classes. In addition to these distinct and focused methods, life cycle analysis and the calculation of the carbon or ecological footprint yield valuable information.

All these methods suffer from the same problem, i.e., some (many?) of the indices are unknown or are estimates, and the multiplication of data errors in any algorithm culminates in large insecurities, making these approaches less useful. At best, these methods reveal differences among groups of pesticides or even among individual compounds. This information might support choices in integrated pest management.

Biodiversity

According to Britannica, "*Biodiversity, also called biological diversity, is the variety of life found in a place on Earth or, often, the total variety of life on Earth. A common measure of this variety, called species richness, is the number of species in an area. Biodiversity also encompasses the genetic variety within each species and the variety of ecosystems that species create.*"

The important aspects of this definition are (1) the distinction between a place or defined location and the Earth as a whole, (2) how it should be measured as the number of species in an area, and (3) the genetic variability within each species.

Similar accessible descriptions, such as https://www.biologicaldiversity.org/programs/biodiversity/elements_of_biodiversity/#:~:text=Greater%20biodiversity%20in%20ecosystems%2C%20species,%2C%20disease%2C%20and%20climate%20change define biodiversity as the variability among living organisms on Earth, including the variability within and between species and within and between ecosystems (Donohue et al., 2016).

All creatures covered by the definition are in some way dependent on each other, either directly (e.g., as food) or indirectly via intermediate creatures. The final view is a pyramid with the most evolved creatures, including humans, on top. Humans dominate this structure because of their cognitive capacities, not particularly because of their physical capacities.

Earth's biodiversity reflects the story of life on Earth with the ongoing appearance and disappearance of species. This story started 4 billion years ago with the appearance of mono-cell-like creatures that have evolved into complex human structures. This evolution was far from linear, with a particularly rapid growth occurring approximately 600 million years ago—called the Cambrian explosion—the period in which most phyla of multicellular organisms appeared, resulting in a marked increase in biodiversity. This growth till today was interrupted five times (the five extinctions): at 480 million years ago (the Ordovician extinction probably due to glaciation), approximately 350 million years ago (the late Devonian extinction due to climate change), approximately 250 million years ago (the Permian–Triassic extinction event probably due to volcanic activity and a decrease in oxygen in the air) killing approximately 96% of all the marine species and an estimated 70% of the land species, called "the great dying," and approximately 200 million years ago (the Triassic-Jurassic extinction) killing approximately 50% of the species (due to underwater volcanic activity and changes in the chemical composition of the oceans). In these four extinction processes, recovery did occur, although the recovery from the Great Dying, after which only a small number of resilient species remained on Earth, took more than 5 million years. Finally, approximately 65 million years ago, the reign of great reptiles, including dinosaurs, ended over a period of thousands of years due to the effects of a collision of the Earth with a large meteor. However, the recovery has become fairly rapid (Donovan et al., 2017), illustrating the difficulty of extrapolation and interpretation of events throughout history.

Most biologists agree that we are in the middle of a new, sixth mass extinction, the Holocene extinction event, caused by ongoing human anthropogenic activity (Jaureguiberry et al., 2022). The discussion on whether the "Anthropocene" is the correct identification of the actual epoch in the Earth's geological timeline or whether it should be replaced by the "pyrocene" is beyond the scope of this book (Pyne, 2024). However, compared to previous extinction events, the changes we have witnessed in the last millennium and even the last decades have occurred at a much faster speed, expressed in tenths of hundreds of years or even decades. This is amply illustrated in Fig. 8.1 (https://www.researchgate.net/figure/Trends-from-1750-to-2010-in-indicators-for-the-structure-and-functioning-of-the-Earth_fig2_370853144).

Adaptability to new environmental challenges still exists, but the capacity to adapt is probably not sufficient to keep up with the fast sequence of changes in this

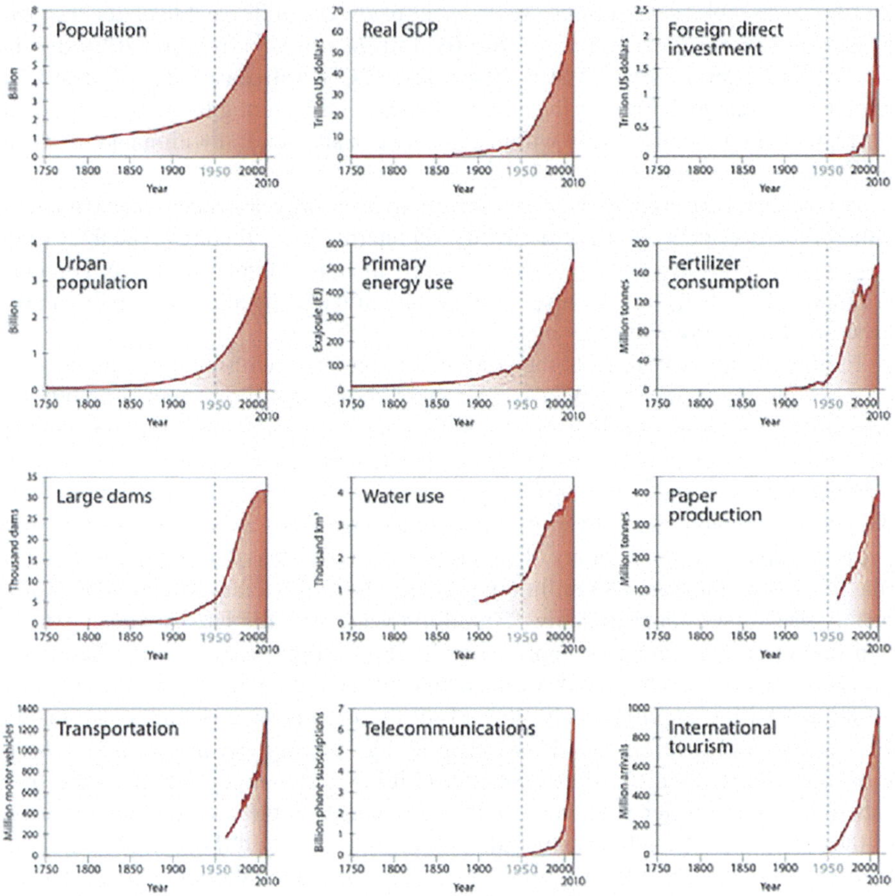

Fig. 8.1 Evolution of societal variables in the function of time (Only part of the variables are shown. Please consult the website for more information)

sixth extinction event. It is impossible to predict further developments over a long(er) period, but important changes are to be expected. The main anthropogenically determined effects are habitat destruction, pollution, and subsequent climate change, which are the main reasons for biodiversity loss (Chapin et al., 2000; Li et al., 2019; Linhares et al., 2023; Rubidge et al., 2012). Under this dooming scenario, most species on Earth will be eliminated within the next century.

The measurement of biodiversity from a human point of view has qualitative and quantitative aspects (Couvet et al., 2011; Lamb et al., 2009; Lindenmayer et al., 2012; Petrou et al., 2015). The qualitative aspect is related to human health and well-being; the quantitative aspect of biodiversity is twofold: the number of species and the total number of living organisms on Earth.

The qualitative relationship between biodiversity and human health is difficult to estimate with some degree of accuracy (Frank, 2024; Irvine et al., 2023; McShane et al., 2011; Naeem et al., 2016; Sandifer et al., 2015; Zhou et al., 2024). The WHO published an extensive report on the relationship between human health and the environment (https://www.who.int/publications/i/item/connecting-global-priorities-biodiversity-and-human-health). At the European level, the impact of biodiversity on human health has been analyzed recently (https://iiasa.ac.at/policy-briefs/mar-2023/future-of-biodiversity-monitoring-in-europe). The importance of biodiversity for health has been stressed on multiple occasions (Atwoli et al., 2021; Chapin et al., 2000; Irvine et al., 2023; Linhares et al., 2023; Robinson et al., 2024; Wildt et al., 2010).

As for the quantitative aspect, recent research is disappointing: we do not know the total number of species on earth. The number of species described in sufficient detail to assign the connotation of species is approximately 2.2 million (figure of 2022), of which approximately one million are insects and approximately 6000 are mammals. The total number of species varies between three million and 100 million, whereas some people estimate the total number of species to be even in the trillion range. Although this is unknown, it is estimated that most of these species are unicellular or of lower cellular organizational structure (https://ourworldindata.org/how-many-species-are-there). The recent worldwide recruitment of scientists to establish a global measurement system is illustrative of this difficulty (Anderson, 2018; Gonzalez et al., 2024; Montoya et al., 2018; Proença et al., 2017; Schmeller et al., 2017, 2018; Spano et al., 2020).

The conclusion is that a definition of biodiversity is available and is comprehensible among all layers of society. Biodiversity on a global scale has changed significantly throughout history, including disappearing due to natural and slowly evolving changes (millions of years) such as ice ages or due to dramatic acute events (probably thousands of years) such as the collision of a giant meteor approximately 65 million years ago. The next question, given the constant disappearance and appearance of species in the Earth's history, is whether the actual situation where an alarming decrease in the diversity of organisms—which is generally accepted—should be taken seriously or is it one of the natural evolutionary episodes in Earth's evolution? The answer tends toward accepting a major human impact on the actual dip in Mondial biodiversity.

The ultimate question is how far biodiversity is indispensable for the survival of humankind or how the variability in organisms in the environment contributes to the survival and wellness of mankind. Ecosystem diversity is crucial to ecosystem integrity, which in turn enables life support, resulting in a livable climate, breathable air, and drinkable water. Food and crop diversity and pollinating insects and bats allow agriculture to support our populations; when a disease affects a food crop, only diversity can save the system from collapse. Greater biodiversity in ecosystems, species, and individuals leads to greater stability in the relationships and interdependency between species. For example, species with high genetic diversity and many populations that are adapted to a wide variety of conditions are more able to resist weather disturbances, disease, and climate change. Greater biodiversity also enriches us with more varieties of foods and medicines.

The importance of preserving biodiversity or so-called “scarce ecosystems” has been abundantly demonstrated (Drupp et al., 2024; Langhammer et al., 2024). The problem, as noted by the authors, is to recalculate the value in dollars of enhanced conservation of the quality environment versus the investment and, more importantly, the return on investment. Currently, it remains very difficult to convince the industry of this value.

Some major conclusions emerge from this short overview: (1) nearly complete extinction is a real concern, (2) extinction can occur over a relatively short period, e.g., thousands of years making evolution difficult to define, and (3) the reappearance of species is possible if enough time is given.

Glyphosate and Biodiversity

Biodiversity is threatened by at least five culprits: climate change, habitat loss, overexploitation, the introduction of alien species, and pollution. Glyphosate at large interferes with two of these five factors. The introduction of genetically modified GLY-resistant crops means, by definition, the introduction of alien species (Schutte et al., 2017), and killing herbs that are not resistant to glyphosate is, again, by definition the reduction in the number of species and hence the reduction in biodiversity. The situation in Europe has been described recently by the European Environmental Agency, which concluded that most of the pressures on nature due to pollution are caused by agriculture (https://www.eea.europa.eu/en/topics/in-depth/biodiversity). Admittedly, killing herbs at one location is not equal to killing herbs all over the planet (see the definition of biodiversity). However, local plants or herbs may be important for local situations. Pollinators depend on flowers; hence, killing flowers significantly reduces the pollinator population, with reduced growth of pollinator-dependent cultures as a logical consequence.

With respect to the chemical aspects of biodiversity, glyphosate is not the most toxic compound, but it is used in such quantities that its ubiquitous exposure compensates for its lesser toxicity (see Chap. 2). It is found in every compartment of the environment, as documented earlier, including in human urine (Connolly et al., 2020; Faniband et al., 2021; Gillezeau et al., 2019; Niemann et al., 2015; Nova et al., 2020), meaning that not only the lower organisms but also the top level of the food chain is affected.

The conclusion is that glyphosate contributes significantly to the deterioration of biodiversity. Notably, glyphosate was originally patented for its chelating properties, and it is a matter of debate whether the effect of glyphosate on the environment could be attributed to its metal-binding capacity or its interaction with other biological systems (Mertens et al., 2018).

The effects on different nontarget organisms were described by Klátyik et al. (2023).

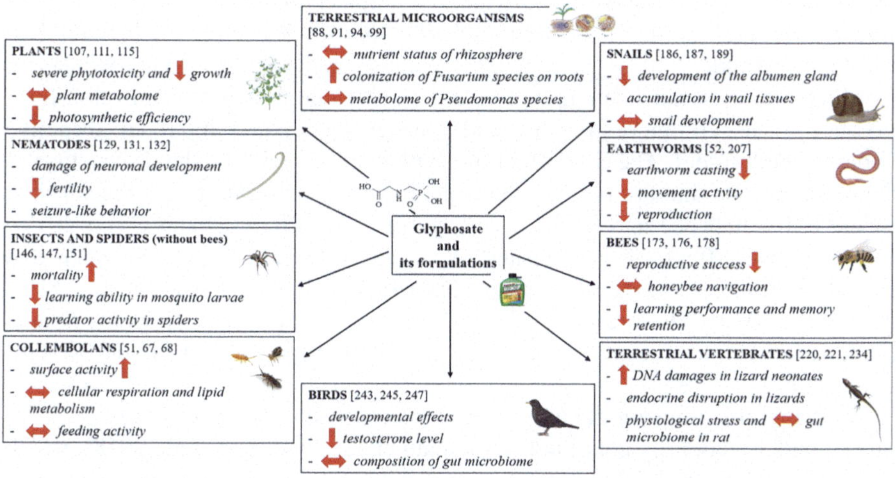

Proponents for further use of glyphosate claim that, since so many chemicals are already involved, some of which are much more toxic than glyphosate, the presence of glyphosate should not be a problem. It is my conviction that this is a reprehensible argument fueled by economic profiteering at the cost of losing the benefits of biodiversity, which in turn is suggestive of complete ignorance and/or deliberate denial of the seriousness of the situation.

There is a consensus that the actual threat to biodiversity is caused by humans, i.e., anthropogenic activities. The concept "Anthropocene" was popularized in 2000 by Stoermer and Crutzen and is defined as the most recent period of less than 1000 years of Earth's history within the Holocene, which started approximately 12,000 years ago (http://people.whitman.edu/~frierspr/Crutzen%20and%20Stoermer%202000%20Anthropocene%20essay.pdf). A full discussion of the concept is not indicated in this text; suffice to refer to a recent book 2021 by Paul Crutzen on the subject (https://link.springer.com/book/10.1007/978-3-030-82202-6).

We can conclude that glyphosate is one of the components that contributes significantly to the ongoing modifications of our ecosystem. Although it might not be the most important culprit, it is likely one of the main components because of its ubiquitous use.

Glyphosate and Pollinators

It is impossible to overestimate the importance of pollinators for the reproduction of plants, flowers, and nearly all food for human consumption. Hence, the need to protect pollinators is imperative, but at the same time, the population has declined by 75% in the last 27 years (https://wikis.ec.europa.eu/display/EUPKH/About+pollinators?preview=/14751600/64651360/Pollinators%20Future%20Brief%20FINAL_WEB.pdf).

Much attention has been given to the domesticated honey bee *Apis mellifera,* because of its commercial value and omnipresence, making it an attractive experimental model. The first effect of glyphosate spraying is the destruction of flowering weeds that are feeding the bees; hence, glyphosate limits the survival of the individual bee and of the hive (Sgargi et al., 2020; Zhang et al., 2019). The negative effect of glyphosate on social interactions and sensory and cognitive abilities has recently been described in detail (Farina et al., 2019; Tan et al., 2022). The presence of glyphosate in honey and its risks for human health are beyond question (Cullen et al., 2019; El Agrebi et al., 2020; Tosi et al., 2022).

Most investigations on the effects of glyphosate on honey bees have been conducted under laboratory conditions (Abraham et al., 2018; Dai et al., 2018), which are different from real field situations. Open field experiments, however, confirmed the deleterious effect of glyphosate on the bee population, such as a significant deformation of the orientation capacity of bees fed low doses of glyphosate/animal. This affects not only the individual bee but also creates a dangerous situation for the survival of the hive (Balbuena et al., 2015; Boksova et al., 2023; Farina et al., 2019; Rampazzo et al., 2023).

The effect on honey bee behavior is illustrated in the next figure (taken from Balbuena et al., 2015).

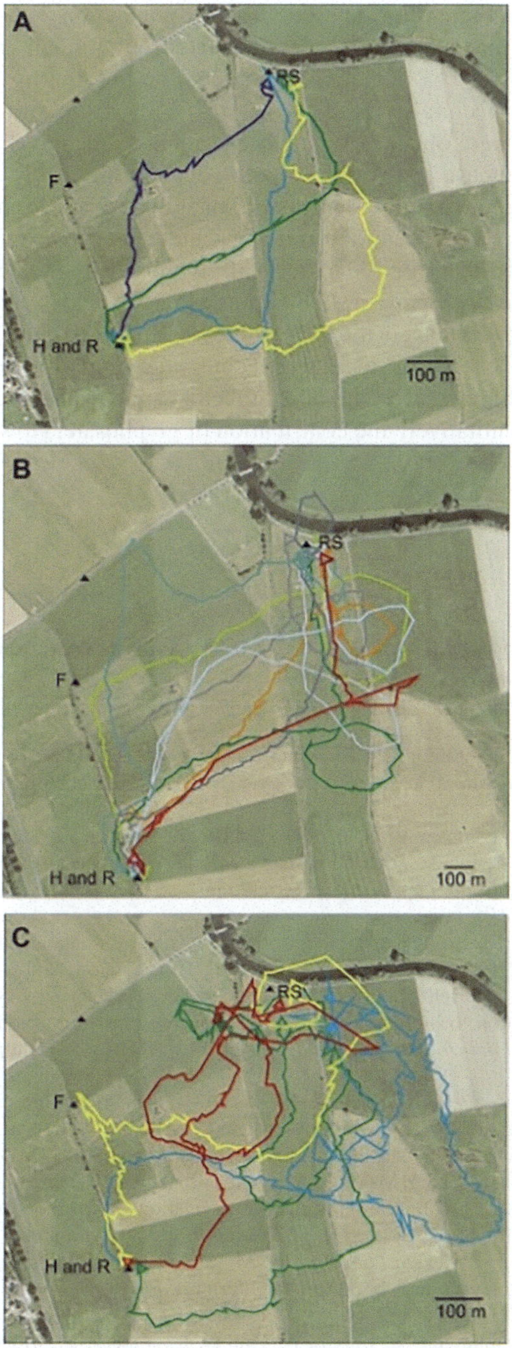

Figure top: Normal flying pattern from the hive to the feeding location. Figure in the middle: Flying pattern after loading the feed with a low concentration of glyphosate (5 mg/l). Pattern low: Flying pattern after loading the feed with a high concentration of glyphosate (10 mg/l)

The bees do not die on the spot, but the total population declines considerably. Glyphosate in the feed causes the bees to deviate from normal, i.e., a trained flight pattern indicating loss of orientation. Some bees do not find hives after being fed high-glyphosate-concentration feed.

Similar results were obtained with other pesticides, such as diquat, leading to a five to ten-fold increase in toxicity to bees (Cech et al., 2022), whereas atrazine caused similar negative effects to those of glyphosate (Wang et al., 2023). Neonicotinoids have been identified as being acutely toxic to bees (Tosi et al., 2022).

As in other discussions on the toxicity of glyphosate, the toxicity of glyphosate for the honey bee is not clear-cut. In addition, other pesticides are more toxic, such as neonicotinoids (forbidden in the EU), whereas the parasitic mite Varroa Destructor is deleterious to bees. It is rational to conclude that glyphosate contributes to the health problems of honey bees, but other effectors are present, and it is hitherto difficult to assign quantitatively the contribution of glyphosate. However, it cannot be neglected.

References

Abraham, J., Benhotons, G. S., Krampah, I., Tagba, J., Amissah, C., & Abraham, J. D. (2018). Commercially formulated glyphosate can kill non-target pollinator bees under laboratory conditions. *Entomologia Experimentalis et Applicata, 166*(8), 695–702. https://doi.org/10.1111/eea.12694

Alister, C., & Kogan, M. (2006). ERI: Environmental risk index. A simple proposal to select agrochemicals for agricultural use. *Crop Protection, 25*(3), 202–211. https://doi.org/10.1016/j.cropro.2005.04.006

Altmanninger, A., Brandmaier, V., Spangl, B., Gruber, E., Takács, E., Mörtl, M., et al. (2023). Glyphosate-based herbicide formulations and their relevant active ingredients affect soil springtails even five months after application. *Agriculture-Basel, 13*(12), 16. https://doi.org/10.3390/agriculture13122260

Anderson, C. B. (2018). Biodiversity monitoring, earth observations and the ecology of scale. *Ecology Letters, 21*(10), 1572–1585. https://doi.org/10.1111/ele.13106

Atwoli, L., Baqui, A. H., Benfield, T., Bosurgi, R., Godlee, F., Hancocks, S., et al. (2021). Call for emergency action to limit global temperature increases, restore biodiversity, and protect health. *New England Journal of Medicine, 385*(12), 1134–1137. https://doi.org/10.1056/NEJMe2113200

Badani, H., Djadouni, F. I., & Haddad, F. Z. (2023). Effects of the herbicide glyphosate n-(phosphonomethyl) glycine on biodiversity and organisms in the soil-a review. *European Journal of Environmental Sciences, 13*(1), 5–14. https://doi.org/10.14712/23361964.2023.1

Balbuena, M. S., Tison, L., Hahn, M. L., Greggers, U., Menzel, R., & Farina, W. M. (2015). Effects of sublethal doses of glyphosate on honeybee navigation. *Journal of Experimental Biology, 218*(17), 2799–2805. https://doi.org/10.1242/jeb.117291

Boksova, A., Kazda, J., Bartoska, J., & Kamler, M. (2023). Effect of glyphosate on the foraging activity of the European honey bee (Apis mellifera L.). *Plant Soil and Environment, 7*. https://doi.org/10.17221/86/2023-pse

Brandmaier, V., Altmanninger, A., Leisch, F., Gruber, E., Takács, E., Mörtl, M., et al. (2023). Glyphosate-based herbicide formulations with greater impact on earthworms and water infiltration than pure glyphosate. *Soil Systems, 7*(3), 18. https://doi.org/10.3390/soilsystems7030066

Cech, R. M., Jovanovic, S., Kegley, S., Hertoge, K., Leisch, F., & Zaller, J. G. (2022). Reducing overall herbicide use may reduce risks to humans but increase toxic loads to honeybees, earthworms and birds. *Environmental Sciences Europe, 34*(1), 20. https://doi.org/10.1186/s12302-022-00622-2

Chapin, F. S., Zavaleta, E. S., Eviner, V. T., Naylor, R. L., Vitousek, P. M., Reynolds, H. L., et al. (2000). Consequences of changing biodiversity. *Nature, 405*(6783), 234–242. https://doi.org/10.1038/35012241

Connolly, A., Coggins, M. A., & Koch, H. M. (2020). Human biomonitoring of glyphosate exposures: State-of-the-art and future research challenges. *Toxics, 8*(3), 18. https://doi.org/10.3390/toxics8030060

Couvet, D., Devictor, V., Jiguet, F., & Julliard, R. (2011). Scientific contributions of extensive biodiversity monitoring. *Comptes Rendus Biologies, 334*(5–6), 370–377. https://doi.org/10.1016/j.crvi.2011.02.007

Cullen, M. G., Thompson, L. J., Carolan, J. C., Stout, J. C., & Stanley, D. A. (2019). Fungicides, herbicides and bees: A systematic review of existing research and methods. *PLoS One, 14*(12), 17. https://doi.org/10.1371/journal.pone.0225743

Dai, P. L., Yan, Z. X., Ma, S. L., Yang, Y., Wang, Q., Hou, C. S., et al. (2018). The herbicide glyphosate negatively affects midgut bacterial communities and survival of honey bee during larvae reared in vitro. *Journal of Agricultural and Food Chemistry, 66*(29), 7786–7793. https://doi.org/10.1021/acs.jafc.8b02212

Donohue, I., Hillebrand, H., Montoya, J. M., Petchey, O. L., Pimm, S. L., Fowler, M. S., et al. (2016). Navigating the complexity of ecological stability. *Ecology Letters, 19*(9), 1172–1185. https://doi.org/10.1111/ele.12648

Donovan, M. P., Iglesias, A., Wilf, P., Labandeira, C. C., & Cúneo, N. R. (2017). Rapid recovery of Patagonian plant-insect associations after the end-Cretaceous extinction. *Nature Ecology & Evolution, 1*(1), 5. https://doi.org/10.1038/s41559-016-0012

Drupp, M. A., Haensel, M. C., Fenichel, E. P., Freeman, M., Gollier, C., Groom, B., et al. (2024). Accounting for the increasing benefits from scarce ecosystems. *Science, 383*(6687), 1062–1064. https://doi.org/10.1126/science.adk2086

El Agrebi, N., Tosi, S., Wilmart, O., Scippo, M. L., de Graaf, D. C., & Saegerman, C. (2020). Honeybee and consumer's exposure and risk characterisation to glyphosate-based herbicide (GBH) and its degradation product (AMPA): Residues in beebread, wax, and honey. *Science of the Total Environment, 704*, 9. https://doi.org/10.1016/j.scitotenv.2019.135312

Faniband, M. H., Noren, E., Littorin, M., & Lindh, C. H. (2021). Human experimental exposure to glyphosate and biomonitoring of young Swedish adults. *International Journal of Hygiene and Environmental Health, 231*, 6. https://doi.org/10.1016/j.ijheh.2020.113657

Farina, W. M., Balbuena, M. S., Herbert, L. T., Gonalons, C. M., & Vazquez, D. E. (2019). Effects of the herbicide glyphosate on honey bee sensory and cognitive abilities: Individual impairments with implications for the hive. *Insects, 10*(10), 16. https://doi.org/10.3390/insects10100354

Frank, E. G. (2024). The economic impacts of ecosystem disruptions: Costs from substituting biological pest control. *Science, 385*(6713), 7. https://doi.org/10.1126/science.adg0344

Gill, J. P. K., Sethi, N., & Mohan, A. (2017). Analysis of the glyphosate herbicide in water, soil and food using derivatising agents. *Environmental Chemistry Letters, 15*(1), 85–100. https://doi.org/10.1007/s10311-016-0585-z

Gillezeau, C., van Gerwen, M., Shaffer, R. M., Rana, I., Zhang, L. P., Sheppard, L., & Taioli, E. (2019). The evidence of human exposure to glyphosate: A review. *Environmental Health, 18*, 14. https://doi.org/10.1186/s12940-018-0435-5

Gonzalez, A., Vihervaara, P., Balvanera, P., Bates, A. E., Bayraktarov, E., Bellingham, P. J., et al. (2024). A global biodiversity observing system to unite monitoring and guide action (AUG, 10.1038/s41559-023-02171-0, 2023). *Nature Ecology & Evolution, 8*(1), 175–175. https://doi.org/10.1038/s41559-023-02263-x

Hagner, M., Mikola, J., Saloniemi, I., Saikkonen, K., & Helander, M. (2019). Effects of a glyphosate-based herbicide on soil animal trophic groups and associated ecosystem functioning in a northern agricultural field. *Scientific Reports, 9*, 13. https://doi.org/10.1038/s41598-019-44988-5

Irvine, K. N., Fisher, J. C., Bentley, P. R., Nawrath, M., Dallimer, M., Austen, G. E., et al. (2023). BIO-WELL: The development and validation of a human wellbeing scale that measures responses to biodiversity. *Journal of Environmental Psychology, 85*, 15. https://doi.org/10.1016/j.jenvp.2022.101921

Jaureguiberry, P., Titeux, N., Wiemers, M., Bowler, D. E., Coscieme, L., Golden, A. S., et al. (2022). The direct drivers of recent global anthropogenic biodiversity loss. *Science Advances, 8*(45), 11. https://doi.org/10.1126/sciadv.abm9982

Kanissery, R., Gairhe, B., Kadyampakeni, D., Batuman, O., & Alferez, F. (2019). Glyphosate: Its environmental persistence and impact on crop health and nutrition. *Plants-Basel, 8*(11), 11. https://doi.org/10.3390/plants8110499

Klátyik, S., Simon, G., Oláh, M., Mesnage, R., Antoniou, M. N., Zaller, J. G., & Székács, A. (2023). Terrestrial ecotoxicity of glyphosate, its formulations, and co-formulants: Evidence from 2010–2023. *Environmental Sciences Europe, 35*(1), 29. https://doi.org/10.1186/s12302-023-00758-9

Kniss, A. R., & Coburn, C. W. (2015). Quantitative evaluation of the environmental impact quotient (EIQ) for comparing herbicides. *PLoS One, 10*(6), 13. https://doi.org/10.1371/journal.pone.0131200

Lamb, E. G., Bayne, E., Holloway, G., Schieck, J., Boutin, S., Herbers, J., & Haughland, D. L. (2009). Indices for monitoring biodiversity change: Are some more effective than others? *Ecological Indicators, 9*(3), 432–444. https://doi.org/10.1016/j.ecolind.2008.06.001

Langhammer, P. F., Bull, J. W., Bicknell, J. E., Oakley, J. L., Brown, M. H., Bruford, M. W., et al. (2024). The positive impact of conservation action. *Science, 384*(6694), 453–458. https://doi.org/10.1126/science.adj6598

Li, X., Wu, J. H., & Li, B. (2019). The battle for biodiversity and human future. *Chinese Science Bulletin-Chinese, 64*(23), 2374–2378. https://doi.org/10.1360/n972019-00283

Lindenmayer, D. B., Gibbons, P., Bourke, M., Burgman, M., Dickman, C. R., Ferrier, S., et al. (2012). Improving biodiversity monitoring. *Austral Ecology, 37*(3), 285–294. https://doi.org/10.1111/j.1442-9993.2011.02314.x

Linhares, Y., Kaganski, A., Agyare, C., Kurnaz, I. A., Neergheen, V., Kolodziejczyk, B., et al. (2023). Biodiversity: The overlooked source of human health. *Trends in Molecular Medicine, 29*(3), 173–187. https://doi.org/10.1016/j.molmed.2022.12.002

McShane, T. O., Hirsch, P. D., Trung, T. C., Songorwa, A. N., Kinzig, A., Monteferri, B., et al. (2011). Hard choices: Making trade-offs between biodiversity conservation and human well-being. *Biological Conservation, 144*(3), 966–972. https://doi.org/10.1016/j.biocon.2010.04.038

Mertens, M., Höss, S., Neumann, G., Afzal, J., & Reichenbecher, W. (2018). Glyphosate, a chelating agent—Relevant for ecological risk assessment? *Environmental Science and Pollution Research, 25*(6), 5298–5317. https://doi.org/10.1007/s11356-017-1080-1

Meys, E. L., Mineau, P., Werts, P., Nelson, S. G. A., Larson, A., & Hutchison, W. D. (2024). Assessment of insecticide risk quantification methods: Introducing the Pesticide Risk Tool and its improvements over the Environmental Impact Quotient. *Journal of Integrated Pest Management, 15*(1), 14. https://doi.org/10.1093/jipm/pmad032

Mohy-Ud-Din, W., Bashir, S., Akhtar, M. J., Asghar, H. M. N., Ghafoor, U., Hussain, M. M., et al. (2024). Glyphosate in the environment: Interactions and fate in complex soil and water settings, and (phyto) remediation strategies. *International Journal of Phytoremediation, 26*(6), 816–837. https://doi.org/10.1080/15226514.2023.2282720

Montoya, J. M., Donohue, I., & Pimm, S. L. (2018). Planetary boundaries for biodiversity: Implausible science, pernicious policies. *Trends in Ecology & Evolution, 33*(2), 71–73. https://doi.org/10.1016/j.tree.2017.10.004

Naeem, S., Chazdon, R., Duffy, J. E., Prager, C., & Worm, B. (2016). Biodiversity and human well-being: An essential link for sustainable development. *Proceedings of the Royal Society B-Biological Sciences, 283*(1844), 10. https://doi.org/10.1098/rspb.2016.2091

Niemann, L., Sieke, C., Pfeil, R., & Solecki, R. (2015). A critical review of glyphosate findings in human urine samples and comparison with the exposure of operators and consumers. *Journal Fur Verbraucherschutz Und Lebensmittelsicherheit-Journal of Consumer Protection and Food Safety, 10*(1), 3–12. https://doi.org/10.1007/s00003-014-0927-3

Nova, P., Calheiros, C. S. C., & Silva, M. (2020). Glyphosate in Portuguese adults – A pilot study. *Environmental Toxicology and Pharmacology, 80*, 6. https://doi.org/10.1016/j.etap.2020.103462

Petrou, Z. I., Manakos, I., & Stathaki, T. (2015). Remote sensing for biodiversity monitoring: A review of methods for biodiversity indicator extraction and assessment of progress towards international targets. *Biodiversity and Conservation, 24*(10), 2333–2363. https://doi.org/10.1007/s10531-015-0947-z

Proença, V., Martin, L. J., Pereira, H. M., Fernandez, M., McRae, L., Belnap, J., et al. (2017). Global biodiversity monitoring: From data sources to Essential Biodiversity Variables. *Biological Conservation, 213*, 256–263. https://doi.org/10.1016/j.biocon.2016.07.014

Pyne, S. (2024). Life in the Pyrocene: Ancient prophecies of worlds destroyed by fire are becoming realities. How will we respond? *Scientific American, 330*(5), 22. https://doi.org/10.1038/scientificamerican0524-22

Rampazzo, G., Gazzotti, T., Zironi, E., & Pagliuca, G. (2023). Glyphosate and glufosinate residues in honey and other hive products. *Food, 12*(6), 13. https://doi.org/10.3390/foods12061155

Robinson, J. M., Breed, A. C., Camargo, A., Redvers, N., & Breed, M. F. (2024). Biodiversity and human health: A scoping review and examples of underrepresented linkages. *Environmental Research, 246*, 17. https://doi.org/10.1016/j.envres.2024.118115

Rubidge, E. M., Burton, A. C., & Vamosi, S. M. (2012). Biodiversity: Past, present and future. *Biology Letters, 8*(1), 3–5. https://doi.org/10.1098/rsbl.2011.0595

Sandifer, P. A., Sutton-Grier, A. E., & Ward, B. P. (2015). Exploring connections among nature, biodiversity, ecosystem services, and human health and well-being: Opportunities to enhance health and biodiversity conservation. *Ecosystem Services, 12*, 1–15. https://doi.org/10.1016/j.ecoser.2014.12.007

Schmeller, D., Böhm, M., Arvanitidis, C., Barber-Meyer, S., Brummitt, N., Chandler, M., et al. (2017). Building capacity in biodiversity monitoring at the global scale. *Biodiversity and Conservation, 26*(12), 2765–2790. https://doi.org/10.1007/s10531-017-1388-7

Schmeller, D. S., Weatherdon, L. V., Loyau, A., Bondeau, A., Brotons, L., Brummitt, N., et al. (2018). A suite of essential biodiversity variables for detecting critical biodiversity change. *Biological Reviews, 93*(1), 55–71. https://doi.org/10.1111/brv.12332

Schutte, G., Eckerstorfer, M., Rastelli, V., Reichenbecher, W., Restrepo-Vassalli, S., Ruohonen-Lehto, M., et al. (2017). Herbicide resistance and biodiversity: Agronomic and environmental aspects of genetically modified herbicide-resistant plants. *Environmental Sciences Europe, 29*. https://doi.org/10.1186/s12302-016-0100-y

Sgargi, D., Adam, B., Budnik, L. T., Dinelli, G., Moldovan, H. R., Perry, M. J., et al. (2020). Protocol for a systematic review and meta-analysis of human exposure to pesticide residues in honey and other bees' products. *Environmental Research, 186*, 6. https://doi.org/10.1016/j.envres.2020.109470

Spano, G., Giannico, V., Elia, M., Bosco, A., Lafortezza, R., & Sanesi, G. (2020). Human health-environment interaction science: An emerging research paradigm. *Science of the Total Environment, 704*, 10. https://doi.org/10.1016/j.scitotenv.2019.135358

Tan, S., Li, G. L., Liu, Z. G., Wang, H. F., Guo, X. Q., & Xu, B. H. (2022). Effects of glyphosate exposure on honeybees. *Environmental Toxicology and Pharmacology, 90*, 9. https://doi.org/10.1016/j.etap.2021.103792

Tosi, S., Sfeir, C., Carnesecchi, E., vanEngelsdorp, D., & Chauzat, M. P. (2022). Lethal, sublethal, and combined effects of pesticides on bees: A meta-analysis and new risk assessment tools. *Science of the Total Environment, 844*, 12. https://doi.org/10.1016/j.scitotenv.2022.156857

Wang, K., Cai, M. Q., Sun, J., Chen, H., Lin, Z. G., Wang, Z., et al. (2023). Atrazine exposure can dysregulate the immune system and increase the susceptibility against pathogens in honeybees in a dose-dependent manner. *Journal of Hazardous Materials, 452*, 10. https://doi.org/10.1016/j.jhazmat.2023.131179

Wildt, D. E., Comizzoli, P., Pukazhenthi, B., & Songsasen, N. (2010). Lessons from biodiversity-the value of nontraditional species to advance reproductive science, conservation, and human health. *Molecular Reproduction and Development, 77*(5), 397–409. https://doi.org/10.1002/mrd.21137

Zhang, L., Rana, I., Shaffer, R. M., Taioli, E., & Sheppard, L. (2019). Exposure to glyphosate-based herbicides and risk for non-Hodgkin lymphoma: A meta-analysis and supporting evidence. *Mutation Research-Reviews in Mutation Research, 781*, 186–206. https://doi.org/10.1016/j.mrrev.2019.02.001

Zhou, Y. C., Huang, Q. X., He, C. Y., Chen, P. Y., Yin, D., Zhou, Y. H., & Bai, Y. S. (2024). A bibliographic review of the relationship between ecosystem services and human well-being. *Environment Development and Sustainability, 28*. https://doi.org/10.1007/s10668-024-04791-3

Chapter 9
EFSA Advice on Glyphosate

The main advising scientific bodies in the EU on the renewal of glyphosate permits are the EFSA and the ECHA, and this advice is published in a number of documents. Three items will be presented in this chapter: (1) an overview of the history of glyphosate permission/renewal processes; (2) the content of the dossier, including critical questions and remarks; (3) some data on the external pressure possibly influencing the decision. This chapter ends with a general conclusion on the EFSA document(s).

Historical Overview of the Glyphosate Renewal Process in the EU

The molecule glyphosate was synthesized in 1950 and patented in 1961 as a metal-chelating agent and antibiotic. The herbicidal activities were detected later, and GLY was marketed in 1974 by Monsanto as "roundup."

In 1996, Monsanto created glyphosate-resistant soya beans by inserting the gene for a modified EPSPS into naïf crops, leading to a culture with a high degree of resistance to GLY. This was patented on 10/7/1990 (https://patentimages.storage.googleapis.com/47/1f/65/9c90643ace1d59/US4940835.pdf). In the years after 1996, other GLY-resistant plants (maize, cotton, canola) became available commercially. Nowadays, more than 9% of the soybeans, corn, cotton, canola, and sugar beets are of the genetically modified glyphosate variant type.

Glyphosate was registered in the EU in 2002 with Germany as rapporteur, according to directive 2001/99/EC. The renewal process was repeated in 2012 by the German *Bundesinstitut für Risokobewertung*, and EFSA/ECHA published their conclusions in 2015, in which the use of GLY was considered to have no health implications. In 2015, the International Agency for Research on Cancer (IARC) published Monograph 112 and concluded that GLY was "probably" carcinogenic to

N. Fraeyman, *Glyphosate 2023–2033*,
https://doi.org/10.1007/978-3-031-97540-0_9

humans (Category 2A). The EFSA was asked to review their conclusions considering the arguments of the IARC. This new report was submitted to the Standing Committee on Plants, Animals, Food and Feed (PAFF), which decided in 2017 that GLY could be further used. Permission was granted for an unusual 5 years until 15/12/2022 as a compromise among the member states that could not agree on 10 years. (https://www.health.belgium.be/sites/default/files/uploads/fields/fpshealth_theme_file/20200303_shc-9561_glyphosate_vweb.pdf).

Owing to the difficulties associated with voting among the member states, the procedure within Europe for permission to introduce a new herbicide or prolong permission was criticized (Frederiks & Wesseler, 2019; Robinson et al., 2020) and in February 2018, the European Parliament set up a special committee, the PEST committee, to review the authorization procedure for pesticides. The committee presented their conclusion in December 2018, which was accepted by the EU parliament by a majority in January 2019 (https://www.europarl.europa.eu/doceo/document/TA-8-2019-0023_EN.pdf). The investigations conducted by the PEST committee were thorough and yielded some significant conclusions and more than 100 recommendations. According to the nongovernmental lobby group Pesticide Action Network (PAN), 15% of the recommendations are implemented, 28% partly or insufficiently, and 57% are not considered, although it is not clear why (www.pan-europe.info/sites/pan-europe.info/files/public/resources/briefings/PANEurope%20PEST%20briefing%2020230425.pdf).

As legally needed, the Glyphosate Renewal Group (GRG), i.e., eight to ten companies led by Bayer, who took over Monsanto on 7/7/2018, submitted a glyphosate dossier for prolongation beyond 2022 before the 3-year deadline. Given the massive amount of information, four European member states were designated for the review (France, the Netherlands, Sweden, and Hungary). The appointment of four member states as rapporteurs instead of one rapporteur and one co-rapporteur was unusual and required an adaptation of the current legal context; this was succeeded with implementing Regulation (EU) 2019/724 and modifying Regulation (EU) 844/2012. These modifications were minor and mostly of legal importance. One particular addition is interesting: *Art1, §1, 5° subparagraph*—"*Prior to the expiry of the deadline for submission of the application, the Member States acting jointly as rapporteur Member State shall agree on the repartition of all tasks and workload.*"

- *From the available information, it is difficult to estimate this repartition. Were the four member states reviewing the whole package of information, or did the reviewing member states partition the workload in fractions? The latter would be logical in view of the appointment of more than one reviewing member state.*

After two consecutive unsuccessful voting attempts, the Commission decided in favor of the prolongation of the use of glyphosate for 10 years from December 2013 till December 2033 (https://eur-lex.europa.eu/eli/reg_impl/2023/2660/oj/eng). Notably, the permission is for the active substance, not for the formulations. It is up to the individual member states to decide on the use of the formulation.

A detailed historical overview of the legal actions related to the earlier and ongoing glyphosate dossier is given in Chap. 11, Annex 4.

On the Content of the Advice

As expected, the amount of information delivered in the renewal process is large. The EFSA has visualized this, as shown in Fig. 9.1 (https://www.efsa.europa.eu/sites/default/files/2023-07/glyphosate_factsheet.pdf).

The application dossier, properly speaking, contains a number of files listed in Table 9.1; the numbering is optional and used for referring purposes in this book only. All these files are freely downloadable at the website of EFSA (https://open.efsa.europa.eu/study-inventory/EFSA-Q-2020-00140).

EFSA has published a number of documents, among them the "renewal assessment report according to 1107/2009 and harmonized labelling 1272/2008" and "the combined draft renewal assessment report prepared according to regulation 1107/2009." The first report is the final report containing the review of all aspects related to health, safety, environmental impact, etc. and the advice on the labeling of the compound upon marketing. The second is the draft version on the aspects of safety, health, and environmental impact.

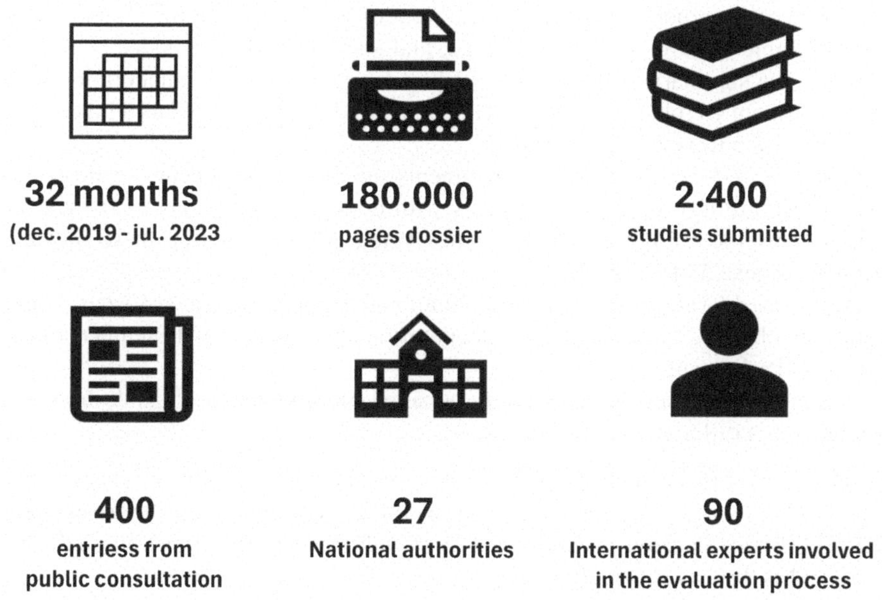

Fig. 9.1 The size of the glyphosate dossier as depicted by EFSA. The internal reference of the glyphosate dossier is *EFSA-Q-2020-00140*

Table 9.1 Supporting documents for EFSA-Q-2020-00140

Nr	Document	Published date
1	Peer review report (61.5 MB)	25/8/2023
2	Glyphosate_AIRV_APPL_zip (4.4 MB)	15/2/2021
3	Glyphosate_AIRV_USSSD_CP_M (22.6 MB)	24/9/2021
4	Glyphosate_AIRV_USSSD_CA_M_7-10 (113.9 MB)	24/9/2021
5	Glyphosate_AIRV_USSSD_CA_M_1-6 (132.8 MB)	24/9/2021
6	Glyphosate_AIRV_USSSD_CP_L (29.9 MB)	24/9/2021
7	Glyphosate_AIRV_USSSD_CA_L_1-5 (102.0 MB)	24/9/2021
8	Glyphosate_AIRV_USSSD_A-I. N (43.7 MB)	24/9/2021
9	Glyphosate_AIRV_USSSD_CA_L_6-10 (106.9 MB)	24/9/2021
10	Glyphosate_additional information request (276.9 KB)	14/3/2022
11	Glyphosate (AIRV) reply additional information request GTF.pdf (91.1 KB)	18/10/2022
12	Consolidated list of newly available publications after commenting period EFSA+AGG March 2023 public.xlsx (143.8 KB)	13/9/2023
13	Glyphosate final RAR public (428.8 MB)	13/9/2023

In this chapter of the book, the document available as supporting documents for EFSA-Q-2020-00140 is used for analyzing the advice of EFSA. Additional information is accessible in the files in the "renewal assessment report according to 1107/2009…"; these are shown in Annex 5 of Chap. 11.

This chapter provides a tiered critical overview at *three levels*. At the *first level*, the "peer review report" of the risk assessment of glyphosate and the second document of Table 9.1 are critically analyzed, and some remarks are added. The peer review is the summary of the advice of EFSA on glyphosate in collaboration with the Assessment Glyphosate Group (AGG) and based on the documents delivered by the Glyphosate Renewal Group (GRG).

At the *second level*, some of the documents supporting the conclusion are evaluated. These are mostly large documents with detailed descriptions of the review processes by the reference member state (RMS). More details on these files are given in Annexe 5 of Chap. 11.

At the *third level*, some of the original animal experiments are analyzed. These results are obtained through juridical intervention upon request of 4 Members of the European Parliament.

The overview is supplemented with personal remarks that are visible from the main text in a different layout (*italics*).

Level 1 This document contains six parts with all comments of meetings with the experts (TC 80, 81 ...), tables etc. It is noted as "comments on the assessment report". This conclusion of EFSA has been published in the EFSA journal; this document is discussed hereafter.

The document was published on 6/6/2023 (put on the website on 25/8/2023):

APPROVED: 6 July 2023
doi: 10.2903/j.efsa.2023.8164

Peer review of the pesticide risk assessment of the active substance glyphosate

European Food Safety Authority (EFSA),
Fernando Álvarez, Maria Arena, Domenica Auteri, Marco Binaglia, Anna Federica Castoldi, Arianna Chiusolo, Federica Crivellente, Mark Egsmose, Gabriella Fait, Franco Ferilli, Varvara Gouliarmou, Laia Herrero Nogareda, Alessio Ippolito, Frederique Istace, Samira Jarrah, Dimitra Kardassi, Aude Kienzler, Anna Lanzoni, Roberto Lava, Alberto Linguadoca, Christopher Lythgo, Iris Mangas, Laura Padovani, Martina Panzarea, Juan Manuel Parra Morte, Simone Rizzuto, Anamarija Romac, Agnès Rortais, Rositsa Serafimova, Rachel Sharp, Csaba Szentes, Andrea Terron, Anne Theobald, Manuela Tiramani, Giorgia Vianello and Laura Villamar-Bouza

This is a 52-page document with five annexes summarizing the results of the risk assessment process (https://efsa.onlinelibrary.wiley.com/doi/pdf/10.2903/j.efsa.2023.8164?fbclid=IwAR1cY_XjOe9G3dBOUSwvHmr3IcamlyrY2LTal21A7k5_BGS-8AWqsX4ahFk).

This is a nice, readable document summarizing the results of all efforts made by the AGG and EFSA. It has 49 references, 6 of them referring to documents published by individual scientists in journals other than the EFSA in-house journal, in OECD guidance documents, in FOCUS guidance documents, or in WHO/JMPR notes. A detailed survey of all the legal steps illustrates the hassle for the applicants. Some critical notes and/or questions:

- On p. 8, it is explained how the literature search is done, and the authors claim that "it is unlikely that relevant papers are missed" based on the fact that the RMS made some critical remarks and because the public consultation yielded some more references.
  - *In my mind, it is the final author(s) who is (are) responsible for the selection of the papers after establishing selection criteria yielding the ultimate list. "Hoping" that no papers have escaped attention is unprofessional.*
- On p. 10, the presence of an impurity that might have clastogenic effects and the lack of relevant information is indicated as a data gap.
  - *The RMS and EFSA could have been stricter on this data gap since clastogenic effects (damage to chromosomes) are important toxicological features. If this impurity—the identity is not revealed!—has clastogenic effects, and if this impurity is present at a sevenfold higher level than proposed, the toxicological consequences should not have been quoted as "the toxicological relevance for this impurity is inconclusive." Precaution could have been a better attitude.*

- The summary of the NOAELs and LOAELs is clear and appreciated.
- On p. 12, reference is made to developmental neurotoxicity studies. It is concluded that glyphosate "does not trigger a concern for parkinsonism" on the basis of human observational studies.
  - *This statement contradicts older (before 2000) and recent (2022–2024) scientific research* (*Bloem & Boonstra*, 2023). *The causal link between glyphosate and autism has not been established, but an indirect effect has been described* in extenso (*He* et al., 2022; *Madani & Carpenter*, 2022). *Whether the neurological effects are due to the trimesium moiety rather than to glyphosate is highly relevant. It is not clear what the basis is of the EFSA advice.*
- P. 14. Many studies have used the commercially available MON 52276 preparation of glyphosate. As I understood, the EFSA should judge the health-related concerns of the active ingredients, but MON 522706 contains at least one additive: Dodigen 4022.
  - *There is no explanation for why MON 52276 is accepted as a reference treatment for animal studies. This is in contrast with the title of the review, which is "…the active substance glyphosate." This is not discussed in the review, and any health effect of the adjuvant is not considered* (*Mesnage* et al., 2021). *It is not clear why this information is not taken into account.*
- On p. 16, "AMPA is a good environmental marker for glyphosate."
  - *I advise the EFSA to reconsider this statement after reading the paper of Schwientek* et al. (2024). *The authors reported no correlation between glyphosate and AMPA concentrations in European rivers, although there is no explanation available. Admittedly, the study of Schwientek needs confirmation.*
- On *p. 17*, the situation with honey is discussed, and increasing the MRL for honey is suggested.
  - *The finding of high(er) glyphosate levels in honey is clear evidence that the honey bee (Apis mellifera) is a preferred consumer and transporter of glyphosate to the hive, although the latter is contested* (*Boksová* et al., 2023). *The causal link between glyphosate exposure and negative effects on pollinators has been established, possibly* via *the microbiome* (*Motta* et al., 2018, 2020), *although this is not the toxic mechanism in the Bumblebee* (*Straw* et al., 2023). *The latter authors found high levels of mortality following contact exposure* (*Straw* et al., 2021), *while the honey bee shows little acute toxic response. A precautionary attitude would be the indicator approach.*
- The chapters on environmental fate and ecotoxicology are clear and comprehensible.
- The endocrine-disrupting properties of glyphosate are discussed on p. 26 of the document. The statement "*However, the criteria for the identification of endocrine disruptors, as laid down in Commission Regulation 2018/605, do not apply*

to formulation" is troublesome. On the one hand, the MON 52276 formulation is used throughout the study; on the other hand, using the formulation prevents conclusions from being drawn. Furthermore, the EFSA uses the EATS system to characterize any endocrine-disrupting effect (see Chap. 4).

- *I assume that EFSA is aware of the paper of La Merrill* et al. (2020) *in which the author proposed ten key characteristics to evaluate endocrine-disrupting effects. The latter reflects more on the complexity of the endocrine system, making it always difficult to establish causal links. Here, again, a precautionary approach would be more appropriate.*

- On p. 31–40, an impressive list of data gaps and outstanding issues is presented. Although most of them are technical and have no deterministic role in the final evaluation the authors conclude: "although not critical, the data gaps may lead to uncertainties in the assessment and are considered relevant."
  - *It is not clear what will happen with this list. The applicant has to respond and fill in the gaps, but when does a gap (and the response to fill the gap) become sufficiently relevant that measures have to be taken? Alternatively, what happens when the answers of the applicants to fill the gaps are not sufficient? Who will judge the response of the applicants?*

Taken together, this peer review document fulfills its purpose of informing the interested reader. Moreover, it contains a number of uncertainties that are worth considering. The overall impression is that whenever a decision in favor or against the further use of glyphosate had to be taken, the balance was mainly, if not always, in favor of the prolongation of the permit and not in favor of possible health concerns. Admittedly, the formal and undeniable link between a particular health problem and glyphosate exposure remains difficult despite the ongoing scientific research yielding results that are highly suggestive of such a link. This is the ideal situation in which to apply the precautionary principle. Notably, the term "precautionary principle" is not mentioned once in the text.

The second document on the list is the "application for approval renewal for glyphosate." This document is submitted by the lead registrant Bayer Agriculture BVBA on behalf of the Glyphosate Renewal Group. This is a 25-page document with information concerning the applicant(s) and information on the chemical identification of the compound. In the appendix, reference is made to "new information." The applicant refers to the previous approval sessions for glyphosate in 2002 and in 2012 during AIR 2. This permit was prolonged in 2017, with an expiration on 15/12/2022. The information delivered now concerns the AIR 5 renewal process. The main point is on the compound used for the dossier. This is "the active substance" (purity of 950 g/kg) and the plant protection product MON 52276.

- *Notably, the MON 52276 formulation contains glyphosate as an isopropyl ammonium salt and Dodigen 4022* (i.e., *a propoxylated quaternary ammonium*) *as an adjuvant. The task of the EFSA is to judge the active substance, so the question is why this formulation is accepted.*

The document contains (Point 4.1) a list of additional studies, of which 2 studies are indicated as "data confidentiality—DC" and 72 studies (of which 1 retracted) are coined with "data protection—DP." Most of these studies are dated before 2020, and the oldest are from 2014. Sections 2 and 4 focus on the properties of the compound and the analytical methods; Section 5 focuses on metabolism, toxico-kinetics, and some revisions of earlier micronucleus tests and AMES tests; Section 6 focuses on the estimation of the residues of glyphosate and AMPA in a series of plants, vegetables, etc.; Section 7 focuses mainly on the behavior of glyphosate in soil; and Section 8 focuses mainly on acute toxicity studies a.o. on bumble bees and solitary bees.

- *These studies are new studies,* i.e., *not submitted and not screened in AIR 2; 2002 and 2015. These studies are assumed to be effective for the evaluation of glyphosate dossier by the Assessment Glyphosate Group (AGG). Notably, no new studies on carcinogenicity (2-year oral feeding studies in rats or mice) are available. This means that all studies on carcinogenicity have been evaluated previously.*

Level 2 Evaluation of the Risk Assessment Report (RAR)

At this level, some RAR documents are summarized and evaluated.

third document, The MCA document coined "glyphosate_AIRV_USSSD_CP_M.zip" contains 12 sections, all of which contain legally required information to be delivered by the applicant. It is the M-CP document (member state concerned position), which reflects concerns about the results presented by the applicant. Any member state can submit (part of) this document. It is all about MON 52276, which is a formulation, not a pure compound.

- **Section 1**: On the identity of the compound (8 pages). The pure compound is described, but there is no information on the composition of the formulation. MON 52276 is a formulation containing a glyphosate salt and an adjuvant.
- **Section 2**: On the physical and chemical properties (19 pages). This section contains data on stability, solubility, etc. Most data come from experiments before 2000. The most recent experiment on the determination of formaldehyde after 2 years of storage is from 2018. This result is negative: The amount of formaldehyde is below the detection limit.
- **Section 3** discusses the possible application of MON 52276 in real life (49 pages). This chapter essentially describes the beneficial effects of glyphosate on agriculture and its implementation in Integrated Weed Management. The avoidance of excessive tillage is proposed as a major advantage.
- **Section 4**: Further information (18 pages). This section describes the precautionary measures to be taken when glyphosate is used, such as the time to wait before harvest after spraying (up to 60 days for vegetables) and the precautionary measures taken by the users (not eating, not drinking, etc.).
- **Section 5**: Analytical methods (54 pages). In this chapter, the analytical methods for the quantification of glyphosate, including the accuracy, reproducibility, and quality of standard curves, are mentioned.

- **Section 6**: Efficacy (5 pages). This is not required for registration of the application. The section is intentionally left blank.
- **Section 7**: Toxicological studies on the formulation (121 pages). This section contains toxicity data, particularly acute toxicity, and no 2-year carcinogenicity studies in rodents. Most of the results are negative: Glyphosate has no or a marginal effect. This is true for acute toxicity (oral, dermal, inhalation, skin irritation, and eye irritation). A skin sensitization test dated 1992 yielded unacceptable results; a new experiment in 2001 demonstrated the absence of skin irritation. A bacterial reverse mutation assay or AMES test (2016) was negative. The in vitro mammalian cell micronucleus assay in human peripheral blood lymphocytes yielded negative results with or without metabolic activation. No clastogenic activity was found when the strains were tested in vitro with human lymphocytes.
- The second main chapter of Section 7 focuses on the exposure of the operator during spraying, bystanders, and residents. To estimate this exposure, the method described by the EFSA is used. (https://efsa.onlinelibrary.wiley.com/doi/epdf/10.2903/j.efsa.2022.7032). Essentially, this method considers the properties of the pesticide (volatility, solubility, etc.) and external conditions during application (wind, temperature, etc.) and locations (vines, orchards, vegetables, etc.). According to the applicants, the exposure in all cases is below the Emergency acceptable operator exposure level (EAOEL) and hence considered without risk. Dermal absorption was studied using human skin samples and C14-labeled glyphosate and was found to be less than 0.1% irrespective of the dose.
- In Table 7-3, it is mentioned that "MON 52276 can be applied safely to operators via tractor-mounted and hand-held application techniques. No specific personal protection equipment (PPE) is necessary," there is no risk for bystanders, no PPE is needed for workers, and no PPE is needed for residents.

  – *However, see Section 4!*

- **Section 8**: Residues in treated products, food or feed (5 pages). Reference is made to the "active substance dossier" while the formulation is studied.
- **Section 9**: Fate and behaviour in the environment (167 pages). Upon the application of 5846.5 g/ha MON 52276, the predicted environmental concentration (PEC_{soil}) was 7.795 mg/kg. The presence of glyphosate in groundwater and sediment is dependent on the soil type, rainfall, and application rate. Figures are given but are highly diverse and difficult to interpret. The main message is that glyphosate is present in soil and groundwater in measurable quantities. Representative figures are given in Tables 9.2.5-122 and 9.2.5-123 after glyphosate was sprayed onto railways track roadsides.

Table 9.2.5-122: $PEC_{sw/sed}$ of glyphosate following application to railways, 1 × 1800 g a.s./ha (HardSPEC 1.4.3.2)

Scenario	Acute (24 hrs) concentration		Application day PEC_{sw} from spray drift (µg/L)
	Water phase (µg/L)	Sediment phase (µg/kg)	
Railway ditch leaching	4.729	16.992	4.729
Railway ditch runoff	4.729	17.000	4.729

Table 9.2.5-123: $PEC_{sw/sed}$ of glyphosate following application to railways, 1 × 3600 g a.s./ha (HardSPEC 1.4.3.2)

Scenario	Acute (24 hrs) concentration		Application day PEC_{sw} from spray drift (µg/L)
	Water phase (µg/L)	Sediment phase (µg/kg)	
Railway ditch leaching	9.458	33.984	9.458
Railway ditch runoff	9.458	34.000	9.458

- **Section 10**: Ecotoxicity (573 pages). The conclusion about the direct toxicity of glyphosate to birds is acceptable: "no visible mortality and long-term impacts on abundance and diversity." Tests were performed with a large range of birds in different locations where glyphosate is applied. The possibility of the intoxication of birds by drinking water was considered negative. The possibility of investigating the indirect effects of birds eating polluted insects was mentioned.
- Acute toxicity to rats, mice, and rabbits is negligible in a Tier 1 approach. The long-term risk was estimated via Tier 2 and 3 approaches, with a similar conclusion: no effect on mortality, abundance, or diversity (Table 10.1.2.35).
- The effects on aquatic organisms were described in detail, confirming the toxicity described during earlier assessments by the EFSA in 2017.

  – *The authors correctly refer to recent papers (Schweizer* et al., 2019) *on the toxicity of glyphosate to zebrafish. This is significant since zebrafish have become an important research tool for a.o. neurological research, cancer research, etc. Recent literature confirms their toxic effects on aquatic organisms (Klátyik* et al., 2023).

The effect on honey bees was analyzed according to the "EFSA guidance document on the risk assessment of plant protection products on bees" of 2013. A low acute oral and contact risk was reported, whereas chronic toxicity (10-day chronic feeding—study conducted in 2017) yields similar results and the conclusion is that glyphosate has no evident toxic phenotype in the honey bee.

– *This conclusion is in contrast with older and ongoing scientific research. Furthermore, the MON 52276 formulation has greater toxicity than the pure glyphosate mixture, which is compatible with earlier remarks that EFSA should evaluate the pure compound and not a formulation.*

The effect on earthworms was not investigated in field experiments because of the calculated low exposure risk provided glyphosate is used as recommended.

- The effects on plants are analyzed in the final part of Section 10. As expected, MON 52276 had significant effects on mortality, shoot height, and shoot fresh weight for all the species tested.
- **Section 11**: Literature data (5 pages). Reference to the literature review report documents.
- **Section 12**: Classification and labelling of the product (5 pages). MON 52276 is considered to have no toxicological hazards, no physical hazards, and no eco-toxicological hazards.

The fourth and fifth documents contain 12 sections; sections 1 to 10 are discussed hereafter. They are MCA documents (member state contribution assessment), which is the peer review of the information submitted by the applicant. It is the country designated to perform the review that submits the MCA document. In the case of glyphosate, four countries contributed. The content of these documents is largely complementary to the documents in the previous list (the position of the MCP concern).

- **Section 1**: Identity of the active substance (12 pages). This contains all the details on the molecule glyphosate. There is no reference to MON 52276 since the additives are considered "confidential." Formaldehyde and the N-nitroso derivative of glyphosate are considered relevant impurities.
- **Section 2**: Physical and chemical properties of the active substance (page 28). Information, such as the melting point, boiling point, vapor pressure, different spectra, and partition coefficient, is given.
- **Section 3**: Further information on the active compound (75 pages). The beneficial role of glyphosate in agriculture is described in detail. Its use and advantages include postharvest for perennial weed control, postharvest for weed control, and pre-planting and post-planting emergence. Application sites are mentioned: orchards and vines, industrial weed control, and railway track vegetation management. The socioeconomic advantages of using glyphosate are described in much detail. The authors noted that no chemical alternative is available and that the nonchemical alternatives have not been adequately compared to glyphosate for efficacy and economic rentability. According to a survey (probably conducted by the applicant), if farmers are deprived of glyphosate, most of them will turn to tillage for weed control. Many figures are mentioned, demonstrating what kind of problems would arise if glyphosate was banned in Germany, Spain, France, and Sweden. The authors launch the term "Conservation Agriculture," and, although it is not clear what this means, glyphosate has to be part of it.
  - *Although the figures are impressive and frequently calculated in many digits, the necessary references are missing. Alternative methods are considered less efficient and are deemed to be subordinate to glyphosate. It gives the impression of a well-orchestrated advertisement campaign.*

The use of glyphosate, e.g., on railway tracks, in particular, some weeds that are traditionally resistant to pesticides (e.g., Japanese knotweed, giant hogweed, and couch grass) or weeds on unsealed surfaces, is promoted. The "smart weed spray concept" refers to the use of digital image analysis technology, leading to localized treatment applications with high efficiency and low off-target contamination. New techniques, such as drones for imagery and drone sprayers and enhanced nozzle control during spraying, would increase the usability of the pesticide.

– *The authors make a great effort to elaborate on techniques that would make the use of glyphosate acceptable. In particular, efforts to avoid contamination are strongly recommended. Why put all this effort into a compound that—according to the authors—is completely safe to handle?*

In the next paragraphs, the authors address the problem of resistance. They claim that in Europe, resistance against many pesticides is already present; hence, glyphosate is the only remaining possibility for weed control. The references are dated from 2004 to 2019.

– *It seems that the authors do not fear the growing resistance of herbs to glyphosate. In their conclusion, "With the exception of two cases reported in wheat, glyphosate resistance cases reported in Europe concern only Conyza and Lolium species and occurred mostly in perennial crops or railways (www.weedscience.org). Most, if not all, of these cases can be attributed to not following label recommendations. The risk of resistance inherent to glyphosate can be considered low." With the current knowledge, this conclusion is wrong.*

- **Section 4**: Analytical methods (721 pages). Table 4.1.1-1 provides an overview of the analytical methods used. More than 200 results are reported, with the oldest report dated 1975 using GLC as the separation method and FPD as the detection method; the most recent report dates from 2021 to the use of LC for separation and MS/MS as the detection method. The methods used in the determination of a large variety of circumstances include, e.g., determination of glyphosate in soil, sediment, and water, as residue in a number of fruits, plants, etc. Analytical methods are also used during toxicity studies in rodents.
  – *The methods are generally well described and executed according to good analytical practice tools. This is not further discussed in this book.*
- **Section 5**: Toxicological and metabolic studies of the active substance (2873 pages). Absorption–distribution–metabolism–excretion (ADME) studies in Sprague–Dawley rats are well documented, revealing an absorption of 30–36% of the dose, a minor metabolism with <1% of the dose identified as AMPA, and excretion mainly via the urine. This pattern is independent of the administration route (oral or IV) and of single or multiple oral doses up to 10 mg/kg. The authors report an in vitro experiment with hepatocytes from rat, dog, mouse, rabbit, and human origin incubated with radiolabeled glyphosate. The minimal metabolism was confirmed.
  – *This experiment with isolated hepatocytes is appreciated, as it confirms the results of the in vivo experiments, whereby all variables are controlled. Also,*

the use of this in vitro methodology will probably contribute to the wanted phasing out of the many animal experiments.

The applicant reports 39 studies on acute toxicity in rats and mice with LD50 values >2000 mg/kg up to 5000 mg/kg. The conclusion is that the compound is not acutely toxic. Acute dermal toxicity was investigated in 27 studies in rats and rabbits. The LD50 values always exceeded the limit dose of 2000 mg/kg; the compound was found to have no acute dermal toxicity. Acute inhalation studies were performed in 22 studies in rats. The LC50 values were above the limit for a dose of 5 mg/L inhaled air. Mortality was observed in 3 out of the 22 studies with rats, while there were multiple clinical signs of intoxication: changes in breathing rhythm, tremors, decreased activity, and dark areas in the lungs. Since these effects were observed only at the highest doses, inhalation toxicity was judged to be minimal.

– *The conclusion of the applicant on inhalation toxicity is possibly not valid for the human situation. The exposure time in these experiments is, as a rule, 4 h, whereas in real life, the exposure periods of farmers or sprayers might be much longer and could be considered (sub)chronic. In this case, the internal exposure level might be an accumulation and subsequent higher internal exposure. There is no information on this point.*

Eye irritation has already been described in previous historical reports. Additional studies confirmed that glyphosate acid or salt is a severe irritant to rabbit eyes and should be labeled as such.

- Twenty skin sensitization tests were performed using glyphosate acid or salt. In none of the studies were any signs of skin sensitization observed.
- For the short-term toxicity–oral 28-day test, the older studies used in previous applications were resubmitted: three rat studies and two dog studies. The argument for not performing additional studies was that enough studies of 90-day oral exposure were available to compensate for the small number of 28-day studies.
  – *As explained by the applicant, only two rat studies were acceptable and resulted in NOELs of 50 and 100 mg/kg. The other studies are, at most, supportive. Studies with dogs are unacceptable. The number of animals was 2 (in the first study) or 1 (in the second study). This does not allow conclusions to be drawn.*

Despite these shortcomings, the applicant concluded that, apart from soft feces and some changes in clinical chemistry parameters, no adverse effects after 28 days of glyphosate administration were observed. This was confirmed by three relevant scientific papers published in 2018 and 2019.

- For the oral 90-day toxicity studies, no new studies were performed, and the older studies used in the application in 2001 and 2015 were resubmitted; nine studies were considered acceptable (in Table 5.3.2-1, ten valid studies are present). The oldest study is from 1978, and the newest is from 1993.

- *Regardless of the results of the studies, the validity of studies that are more than 30 years old should be questioned. The first OECD guidelines for animal toxicity testing were published in 1981 and were expanded considerably in the 1990s. It is assumed that experiments in the nineties were not fully compatible with these guidelines.*

These studies revealed that the lowest NOAEL values were 79 and 90 mg/kg for male and female rats, respectively.

- *The limited quality of the studies is apparent when comparing NOAELs that ranged from 79/90 to 1267/1623 mg/kg for males and females, respectively. This range is too large for conclusions. Nevertheless, the applicants claimed low toxicity in the 90-day toxicity studies.*

The information on genotoxicity and carcinogenicity is presented in detail in approximately 700 pages, reflecting the importance of the ongoing discussion on this item. The overall standpoint of the applicant is that glyphosate is not genotoxic. The applicant considered the numerous publications demonstrating DNA damage unsuitable for several reasons: inadequate identification of the glyphosate source and inconsistent methodology (a.o., cell lines, exposure time, etc.). The authors conclude that cellular toxicity is the prime cause of DNA damage and that glyphosate does not interact directly with DNA.

- *The use of the TA 98 and TA 100 cell lines in the AMES test is complemented with TA 102, TA 1535, and TA 1537. The latter strains allow the determination of different DNA damage processes; TA 102 allows the detection of oxidative stress (ROS generation) as a possible cause of DNA damage. Notably, a number of earlier studies (<2000) performed according to the OECD guidelines of 1983 and submitted in 2015 were considered invalid for the RAR (lacking TA 102 and using impure glyphosate test material). Nevertheless, these studies were resubmitted and considered supportive.*

Twenty-four results of the AMES test are available, of which 12 were considered valid according to the OECD guidelines and 12 were considered "supportive." All tests were negative. Twelve tests were performed before 2000, the others were more recent, the youngest test from 2014. On the basis of their observations, the applicant concludes, "*On the basis of the experimental findings, the test item is not mutagenic in bacteria with and without metabolic activation.*"

- *A negative AMES test does not necessarily mean that a compound is not mutagenic. Furthermore, the possibility that glyphosate has an indirect effect on DNA via other cellular toxicity mechanisms is an important suggestion that has already been mentioned in several publications.*

Chromosome aberration tests have been performed with Chinese hamster lung and ovary cells, human peripheral lymphocytes, mouse lymphoma L5178Y cells, HGPRT assays in CHO cells (1983), HPRT assays in V79 cells (2021), and micronucleus tests in human lymphocytes (2021). The tests of 2021 were not reported in earlier applications. All tests were negative except for the two HPRT tests of 2021, which scored a reduction of approximately 30% in mitogenic activity.

– *A 37% and 33% reduction in mitotic activity for the two tests with human peripheral lymphocytes is considered irrelevant. However, this reduction means that the test substance has some effect on cellular activity, although the precise target is not defined. This result cannot be denied.*

Seven studies on DNA damage and repair are submitted, although they are considered invalid in the discussion on glyphosate genotoxicity.

– *The number of studies and the amount of information delivered by the applicant are impressive. However, many of them were considered "not valid," "not accepted," or at best "supportive." This inevitably contributes to the massive number of pages in the application. It might be worthwhile to restrict the application to tests that are effectively used, valid, and acceptable.*

Thirteen in vivo micronucleus bone marrow tests were presented: 1 in rats and 12 in mice; 9 studies were considered valid. All studies were negative. Two groups of studies were distinguished: 8 studies were from before 2000, 7 studies were performed after 2006–2007, most of the older studies were either invalid or supportive, glyphosate was administered by gavage on 2 consecutive days, and the tests were performed 1 day later. All studies (valid or not) yielded negative results, and the applicant concluded that glyphosate has no potential to induce micronuclei in the bone marrow of rats or mice.

– *Here again, the experiment with male Swiss mice is reported in extenso, while the study is irrelevant because the dose levels are far too low to detect any effect. This is not very useful. Moreover, a micronucleus study in which the mice were treated via the intraperitoneal route was performed. Since the whole dossier aims at discussing human/environmental health, it is peculiar why this study is included mainly because the applicant concludes "…this is not representative of human exposure."*

Three dominant lethal tests in rats and mice demonstrated the absence of any effect of glyphosate on germ cells. A significant reduction in the number of pregnant females after the first mating was found, together with evidence for pre- and post-implantation losses. The applicant concluded that glyphosate has no dominant lethal effects in Wistar rats and is considered nongenotoxic to germ cells in vivo.

– *None of the studies are valid for several reasons: The number of animals is too low, and the number of implants is less than the recommended number (400). The results are at best, considered supportive. References to several papers that are less than 10 years old are presented and considered supportive of the conclusion of the applicant. These papers are discussed in detail in their text.*

Of major importance are the long-term toxicity and carcinogenicity studies. Nine studies with Wistar and Sprague–Dawley rats are presented in Table 5.5-1, including seven 2-year studies, one 26-month study, and one 1-year study. The NOAEL and LOAEL are presented in Fig. 5.5.1--1; the overall NOAEL is 300 mg/kg.

All studies, except a study from 1981, are considered valid. As an example, the values for NOAEL and LOAEL are shown in Fig. 9.2.

- Data on body and organ weights upon necroscopy, data on clinical chemistry, and data on nonneoplastic and neoplastic lesions are given and suggest the non-toxic effects of glyphosate upon chronic administration. Some lesions are frequently found (e.g., fibroadenoma in female rats and benign neoplasms in male and female rats in the 1997 study); however, these lesions are due to age and independent of glyphosate treatment. In a 1996 study (Table 5.5-55), an increased occurrence of pheochromocytoma tumors in treated male rats was observed, whereas in a subsequent 1997 study, pheochromocytomas were found only in nontreated animals (see further). Adenomas are frequently found, but this is most likely due to age and not treatment.
- Studies in mice have shown comparable results. Of the eight studies, three studies were performed after 2000, five studies are older, and the oldest is from 1982. The overall result in mice was that no neoplastic lesions could be attributed to the treatment; if they were present, they were due to the normal age-related appearance of the lesions.
- In a separate chapter in Section 5, the applicant discusses numerous papers with epidemiological findings on the possible carcinogenicity of glyphosate in humans. The claim of IARC (glyphosate is "probably" carcinogenic to humans) takes a central position in this discussion. Papers that are considered reliable or nonreliable are presented at length, summarized, and evaluated. For a short overview of the essentials, please refer to Chap. 3 of this book.
- The next chapter focuses on generational studies. According to the applicant, there are no effects on the offspring (first- and second-generation) of dams

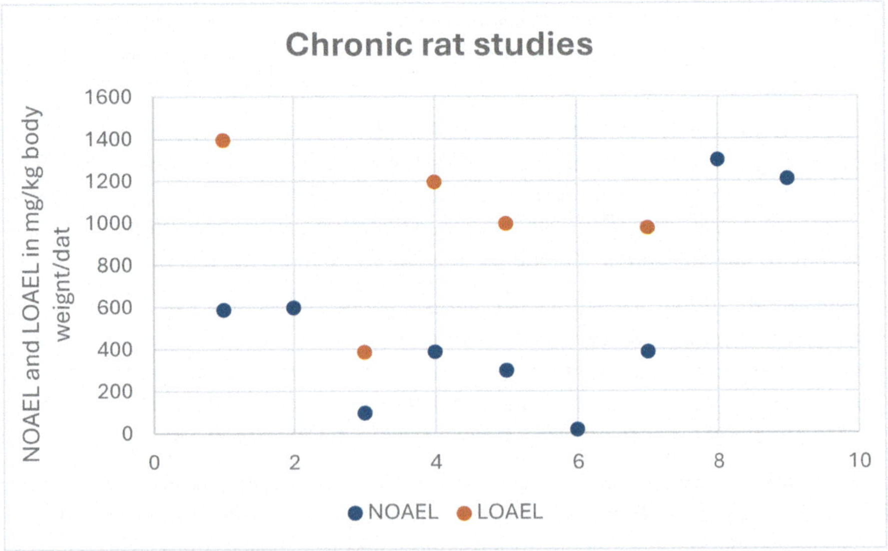

Fig. 9.2 NOAEL and LOAEL obtained during the 2-year carcinogenicity studies with rats

treated with glyphosate via the diet. In addition, 11 most recent papers are presented, mostly demonstrating no generational effect. Here, again, most studies were performed before 2000, the oldest one was from 1981, and two studies were more recent (2000 and 2007).

- Developmental effects, i.e., toxicity to mouse dams and offspring, were investigated in seven studies. No effect on the properties of the offspring (number, sex ratio, percentage resorption, fetal death, or fetal weight) or the dams (weight and weight gain) was detected, except for minor gastrointestinal disturbances. The applicant concludes that glyphosate has no teratogenic properties. Similarly, the results of the rabbit studies (three valid, two supportive) indicate that toxicity to fetuses is due to maternal toxicity without teratogenicity. In one study, all rabbits treated at 1000 mg/kg died on day 29 after administration. The usefulness of rabbit studies for the evaluation of the human health situation is questioned.
  - *Although the number of papers in this and previous chapters is high, some bias is possible. Additionally, some papers are missing, e.g., the paper by Kubsad* et al. (2019)*, which demonstrated the absence of any effect in the F1 generation, but the F2 and F3 generations presented increased health problems such as prostate, ovarian, and kidney diseases and obesity* (*Kubsad* et al., 2019).

The applicant presents studies on neurotoxicity and concludes that in rats and domestic chickens, no evidence for acute or delayed neurotoxicity was found. Neurotoxicity was assayed with a battery of functional and behavioral tests. The applicant stresses that glyphosate is not an inhibitor of acetylcholinesterase, which is generally known for its neurotoxic effects.

- *The authors take great care in citing literature data and obviously—they must know the paper by Samsel and Seneff* (2013)*. See Chap. 5 for commentary. Although the Samsel–Seneff paper is wrong, it would fortify the belief and acceptance of the literature reviews of the applicant when taken up as a reference.*

The next chapter discusses the toxicity of the metabolites, in practice, of AMPA. Overall, AMPA has no acute or chronic toxic effects.

- Some additional studies that are legally not required in the dossier are presented, including a study on immunotoxicity in female mice. The spleen cell number and IgM antibody-forming cell response were unaltered, indicating the absence of immunotoxicity. Other studies involved (1) effects on the salivary gland; (2) effects on behavior, cardiac function, and muscle contractility in rats; (3) coadministration of other pesticides, such as 2-4D; (4) acute oral toxicity in female goats and heifers; (5) irritating effects on the gastrointestinal mucosa of anesthetized dogs; (6) comparisons of toxicity after IP or oral administration; and (7) dermal absorption through damaged rabbit skin. In none of these studies were meaningful conclusions possible; these studies are not discussed in this book.
  - *The efforts of the applicant to deliver additional information that is not legally demanded are appreciated. The study of immunotoxicity is important since*

damaging resistance directly or indirectly has important health consequences. The authors should refer to recent papers on the subject, e.g., *Peillex and Pelletier* (2020).

The following chapter discusses the possible endocrine-disrupting effects of glyphosate. The conclusions of the applicant are based on ten studies on EATS-mediated endocrine-disrupting properties, all of which are considered valid. In short, there was no androgen or estrogen receptor binding activity, glyphosate has no estrogen agonist activity or aromatase inhibitory activity, there was no uterotrophic activity or activity according to the Herschberger assay measuring androgenic activity in castrated rats and no evidence for puberal thyroid development in male and female rats, and the non-inhibitory character of glyphosate in the H259R steroidogenesis test was demonstrated. Taken together, the applicants have no evidence that glyphosate has endocrine-disrupting properties.

- Two additional studies are presented. In the first study, the OECD H259R standardized steroidogenesis assay protocol was validated via the use of 28 chemicals as model substances, one of which is glyphosate. Seven laboratories participated, and two laboratories investigated glyphosate. Glyphosate scored negative in this multicenter experiment. The second study was initiated following the recent EU guidelines on endocrine-disrupting agents, i.e., the necessity of a demonstrable link between the substance and the effect. The applicant concludes that the requested link was not demonstrable; hence, according to EU guidance, glyphosate has no endocrine-disrupting properties. This conclusion is further supported by a selection of seven recent scientific papers (2016–2020).
  - *According to the EU rules, the link between cause (the disruption of endocrine functionality by a compound) and consequence (a phenotypically recognizable health problem) must be established beyond doubt, and if not, the compound is not endocrine-disrupting. The complexity of the endocrine system is such that this link is very difficult to pinpoint. This phenomenon was well understood by La Merrill* et al. (2020), *who described ten key characteristics of compounds with endocrine-disrupting properties. The applicants should confront their results with these characteristics. This has been done (see Chapter on endocrinology) and glyphosate is correlated with eight out of the ten key characteristics. See Chap. 4 of this book.*

The final chapter in Section 5 focuses on some medical characteristics of glyphosate, including the protective measures taken by those handling the compound. Of interest are the results of their first study on the glyphosate concentration in dust on the premises of a glyphosate-producing company and the corresponding exposure of workers over a period of 30 years. Workers are exposed to glyphosate at concentrations between 1 and 10 μg/kg/day. No medical data of the population over that period are given. In a second study, the urine of horticulturists spraying glyphosate was analyzed post-spraying, and the calculated half-life was estimated at 4.5 (unadjusted) or 7.5 (creatinine-adjusted) hours.

  - *The applicant is obviously aware of the large number of recent papers on the presence of glyphosate in human urine as well as in many other environmental compartments. The references in their document are only a small selection.*

- **Section 6**: Residues in or on treated products, food and feed (1250 pages).
- The majority of this section focuses on analytical methods applied to a large variety of crops, vegetables, and animals used for meat production. As mentioned in the introduction, this will not be discussed. The methods are generally well described and performed according to good analytical practices (standard curves, blanks, positive and negative controls, in some cases, multicenter analysis). This is considered correct.
- In the second part of this section, the residues in food and feed are described. The applicant distinguishes between the crops genetically modified to resist glyphosate treatment and others, but this is not applicable in the EU. From the concentrations of glyphosate and AMPA, the MRLs are derived, leading to Table 6.9--2 with the proposed MRLs. The values for fruits, vegetables, oil seeds, teas, coffee, herbal infusions, and sugar plants are all set to 0.05 mg/kg. For products of animal origin, the values range from 0.06 to 0.15 in muscle and fat tissue; 0.5 in bovine, sheep, and goat liver; and 0.7 in swine and equine kidney. Under the assumption that the ADI and ARfD are 0.5 mg/kg/day, no toxicity is expected. Furthermore, the applicant refers to papers confirming their proposition. One paper focused on the presence of glyphosate in baby food and concluded that "*baby food commodities indicate compliance with existing legislation.*"
- In a separate study, the presence of glyphosate and AMPA in honey was investigated. After a single application of glyphosate MON52276 in four semi-field trials at different locations, the glyphosate concentration in the honey varied between 0.87 and 6.9 mg/kg, which resulted in an average MRL of the four trials of 1.5 mg/kg. These experimental data are supplemented with the data in Table 6.10.1-6, in which all estimations (approximately 400) of the residue levels of glyphosate in honey from 2016 to 2017 are presented. Most of the values are between 0.01 and 0.03 mg/kg, with some outliers of 0.1 and 0.4 mg/kg.
  - *The difference between the experimental data and the real-life data is noteworthy and might reflect the overshoot of glyphosate exposure during the experiment. It is not possible to control this point.*

Finally, the applicant discusses four additional scientific papers confirming their findings: The concentration of glyphosate in honey is below the MRL established by the EU (0.05 mg/kg). The final part of the section focuses on an additional analytical method (ELISA) for glyphosate determination in honey and on geospatial analysis when honey is collected. Neither item is discussed in this book.

- **Section 7**: Fate and behaviour in the environment (2564 pages).
- Here, again, as in Section 6, the main focus is on the analytical methodology; this will not be discussed.

- Aerobic soil metabolism under laboratory conditions revealed that glyphosate was rapidly degraded to AMPA and ultimately to CO_2 depending on the incubation time (from 60 to 364 days). When C^{14}-glyphosate was used, some unknown radioactive contaminants were detected, although they were always present in minor quantities. The results of degradation under anaerobic conditions were similar. The differences among the different soil types are noted, but the overall conclusion is the same: Glyphosate is degraded to CO_2.
  - *In the series of experiments on aerobic degradation, one is from 2010, and the remainder is older than 30 up to 52 years! The results have been presented several times during the sequential applications of the pesticide for renewal of the approval. This is not new information. Studies on anaerobic metabolism are more recent, with the youngest being from 2004.*

The degradation rate is dependent on the soil type, acidity, and temperature and is 50% between 30 and 105 days and 90% degradation between 100 and 3450 days.

- The soil dissipation studies revealed that glyphosate and its degradation products were present in the upper 10 cm of the soil, depending on the type of soil. Three studies were performed in 2020, and the remainder of the results were at least 30 years old.
- Studies related to absorption, desorption, and mobility in soil have been performed; however, they are not discussed here.
  - *These studies were performed with the trimesium salt glyphosate. Notably, this salt form is no longer in use in the EU because of the toxicity of the counterion.*

Lysimeter studies have demonstrated that both glyphosate and AMPA, when present in the soil, can leak into groundwater, whereas Bergström et al. reported that this possibility is considered low under European climatological conditions (Bergström et al., 2011).

- Glyphosate in water is stable and largely independent of pH; less than 10% of glyphosate is hydrolyzed within 5 days. In addition to being relatively stable in aquatic environments, glyphosate is also resistant to biodegradation; approximately 26% of the mother molecule in the presence of nitrogen was transformed into ammonium salt after 28 days.
- The half-life of glyphosate in the air was 0.135 days (1.6 h), which is indicative of fast degradation. The main reaction is between glyphosate and photochemically produced OH− radicals; the role of ozone is minimal since glyphosate does not contain olefinic or acetylenic sites. Furthermore, the amount of glyphosate evaporating from plant surfaces or soil is 20% at most of the applied doses of glyphosate.

– *The stability of glyphosate in the air is important because of the potential exposure of bystanders during spraying.*

In a final overview, the applicant summarized all the information gathered in the context of Section 7. Here, again, the main focus is on analytical methods for the determination of glyphosate under diverse circumstances and environmental conditions. The applicant refers to 75 peer-reviewed scientific reports largely supporting the data.

- **Section 8**: Ecotoxicological studies on the active substance (909 pages).
- This section is compatible with Section 10 of the CA-M document. A short overview is given here.
- Sixteen studies on the acute and 8-day treatment toxicity of glyphosate and AMPA to birds were presented, the most recent of which was dated to 2003, and the oldest was from 1983. The conclusion is that glyphosate is not acutely toxic to birds. There was no effect on reproduction during the 20–21-week feeding study. No bioaccumulation is expected since the log octanol/water coefficient is >3.
- Endocrine-disrupting properties were not observed according to the EATS guidance. The toxicity to fish is well known and was confirmed in 20 studies involving acute exposure and four studies involving chronic exposure. All these studies support the conclusion that glyphosate is toxic to aquatic organisms, including *Daphnia magna*.
- The acute toxic effect and the effect of contact toxicity of glyphosate on bees (*Apis* mellifera) were studied in 16 studies. The conclusion was that the LD50 value was >104 µg/bee. One study in 2017 was performed on a chronic (10 days of oral feeding) toxicity test with five concentrations of glyphosate in a sugar solution. The LD50 value was >179 µg/bee. No differences from the control treatment were found for the survival of larvae upon exposure.
- For the toxicity to nontarget arthropods, earthworms, nontarget higher plants, and other terrestrial organisms, please see Section 10 of the CA-M document.
- **Section 9**: Literature data (10 pages)
- This section does not provide the reader with the results of the literature search but instead describes the method for performing the search. The period of 10 years prior to the submission of the dossier was covered (from January 2010 to June 2020); 690 papers were considered sufficiently relevant for reporting purposes.
- **Section 10**: Classification and labelling of the active substance (7 pages).
- The applicant proposes the following label for glyphosate. A safety data sheet is presented.

Pictograms:	
Signal words:	Danger, GHS09, GHS05
Hazard statements:	H318: Causes serious eye damage
	H411: Toxic to aquatic life with long-lasting effects
Precautionary statements:	**Prevention**
	P273: Avoid release to the environment.
	P391 Collect spillage.
	Response
	-
	Storage
	-
	Disposal
	P501 Dispose of contents/container to an installation for the handling of hazardous waste approved by the competent authority

The role of EFSA and ECHA is clear and different: EFSA advises on the risk assessment after exposure, whereas ECHA is concerned with the intrinsic toxic properties of the compound and suggests a rule for the classification and labeling regulation of the compound. The EFSA draws conclusions on the Renewal Assessment Report (RAR), which is written by the rapporteur member state, in the case of glyphosate in the 2023 renewal process: four countries. ECHA has an internal Committee for Risk Assessment (RAC) that is responsible for the ECHA advice.

Both assignments are very congruent. The last sentence of the abstract of the paper "conclusion on the peer review of the pesticide risk assessment of the active substance glyphosate" (https://doi.org/10.2903/j.efsa.2015.4302), published in the EFSA Journal in 2015, reads as follows: "*Following a second mandate from the European Commission to consider the findings from the International Agency for Research on Cancer (IARC) regarding the potential carcinogenicity of glyphosate or glyphosate-containing plant protection products in the on-going peer review of the active substance, EFSA concluded that glyphosate is unlikely to pose a carcinogenic hazard to humans and the evidence does not support classification with regard to its carcinogenic potential according to Regulation (EC) No 1272/2008.*"

This is clear and unequivocally proof that the classification is the assignment of the ECHA. This assignment was defined by the ECHA as follows: "*Harmonized classification and labelling focuses only on the hazardous properties of the substance: its potential to cause harm. It does not assess the exposure of people or the environment to glyphosate. This will be part of the peer review of the risk assessment done by EFSA*" (https://echa.europa.eu/nl/-/glyphosate-no-change-proposed-to-hazard-classification).

Confusion from this double regulatory approach is possible, although both instances usually reach similar conclusions.

The most important overall remarks are summarized hereafter.

- The beneficial effects of glyphosate for agriculture are exposed in extenso. The applicants introduced the term "conservation agriculture," which includes minimal soil disturbance (no tillage), permanent soil organic cover, and species diversification (https://www.fao.org/conservation-agriculture/en/). According to the authors, glyphosate is an essential ingredient since glyphosate prevents tillage. Another item is the "smart weed spray concept," which indicates the use of artificial intelligence, advanced detection methods for the presence of herbs, and precision spraying with new devices such as drones that eliminate problems with glyphosate. There is not any reference for the possible negative effects of glyphosate. It seems contradictory that much protective effort is proposed vis-à-vis to what is considered by the RMS and EFSA as innocent molecules.
- It is unlikely that the authors are not aware of the increasing resistance of herbs to glyphosate. Many recent papers are quite clear on this point, and the biological mechanisms are described (Hulme, 2023; Ofosu et al., 2023). Although plant resistance by itself is not a threat to human health, it will force farmers to use either other (possibly more toxic) compounds or increase the dose of glyphosate. Both are to be avoided.
- The tests for acute respiratory toxicity are not compatible with the use of glyphosate in real life since farmers and sprayers work for longer than 4 h in a row. Agriculture is not a 9 to 5 job, and the tests do not mimic this situation. The results are unreliable.
- One important remark that is valid throughout the whole document is that most studies are old (more than 25 years), and some of them are very old (up to 50 years). These studies have been presented in previous revision rounds at the EU level (2002 and 2015) and have been accepted. It would be strange that conclusions opposite to the conclusions of 2002 and 2015 would emerge now.
- For most parts (analysis, methods, etc.) that are not directly involved in human health, the description and evaluation are correct. For some others (e.g., acute pulmonary exposure), better experiments are to be done, and for others (e.g., carcinogenicity studies), old and very old studies are presented. The results of new studies are wanted.

Level 3 Analysis of individual animal experimental data

In 2018, four members of the European Parliament Green Party requested public access to glyphosate files on carcinogenicity. The EFSA refused to deliver, data protection as the argument. The European court refuted the argument of EFSA and ordered the publication of the files. Consequently, 16 files (further called "public files") were released: nine studies involving 2 years of feeding or different rat strains and five studies involving 2 years of feeding of mice. The administrative information of the studies is summarized in Chap. 11, Annex 6.

The first question is whether these studies are part of the studies used for the evaluation of the carcinogenicity potential of glyphosate. The codes/identification tags of the studies were compared with the data in the RAR and in the files of the

GRG, and while for the rat studies, the data from the RAR and the GRG are compatible, five out of the nine rat studies from the public files were detected in the RAR and GRG; see the table in Annex 6. This means that the EFSA has made public four rat files that were not reported by the GRG within the context of the application. It is not evident where these four study results were derived from. For the mouse data, all the data in the public files are detected in the RAR and GRG documents.

The second question concerns how these public files are interpreted by the RAR and to what extent this interpretation is subject to criticism. A complete analysis of these public files was performed, and the main observations and conclusions are given hereafter.

- Major differences between treated animals and controls are found without any explanation. For example, in Study 1, pheochromocytomas were found in treated male rats but not in controls. In Study 4, pheochromocytoma tumors were found in control animals but occurred less frequently in treated animals. The occurrence of pheochromocytomas was also noticed in the 1996 study (Table 5.5-55), Section 5 in the review of the fourth document.
- The survival of these animals is troublesome. In Studies 2, 4, 6, and 7 with rats, the animals treated with the highest dose had longer survival than the control animals did. In Study 10, this was the case with mice. In the other studies, this is not the case. In no instance is any explanation or concern given.
- As far as neoplastic lesions are concerned, in all studies, the conclusion of the authors is that glyphosate has no carcinogenic properties, although the incidence of tumors is very high in some studies. For example, in Studies 2, 4, 6, 7, and 11, the occurrence of tumors in treated animals was greater than in control animals, while in other studies, the reverse was true (e.g., Studies 1 and 12), which makes—by definition—any conclusion doubtful. The explanation of the authors is variable. In Studies 5, 9, 10, and 14, the occurrence of tumors in treated animals is said to be WITHIN HISTORICAL DATA for this type of animal and therefore not important. In other studies, both arguments (i.e., the absence of difference between control and treated animals and the comparability with historical data) are used simultaneously to explain the findings (e.g., study number 10). In Study 9, the occurrence of tumors in treated animals was not considered important since it was not dose-dependent and was not significantly different from the controls.
- In some cases, this conclusion of the authors can be questioned. In contrast to the point of view of the authors, in Study 1, the occurrence of pituitary adenomas in female rats and pheochromocytoma in male rats could be related to treatment. In Study 5, pancreatic adenoma cell tumors and in Study 13, lymphoid malignant tumors could equally well be described as a treatment-related effect. In Study 9, a clear increase in the number of tumors in the high-dose male and female groups was observed, which was not detected in the control animals. In this case, the absence of statistical significance and dose dependency was invoked as the explanation for non-carcinogenicity.

Taken together, the results of these studies are difficult to interpret. The authors conclude that glyphosate has no carcinogenic properties. Some studies on the occurrence of tumors in treated animals are supposedly related to the high incidence of tumors in control animals, but this explanation is not always feasible. The possibility that glyphosate has some carcinogenic effects cannot be excluded.

The Scientific Literature on Glyphosate, the Practice of Lobbying

In the following paragraphs, I will first briefly discuss the information delivered by EFSA and the AGG on the scientific literature and, secondly, comment on the interference of groups displaying fraud, self-interest, financial and/or economic interests, lobbying groups, etc.

EFSA and the AGG mentioned the exceptionally high number of (scientific) papers delivered by the applicant. Upfront, the GRG excluded a number of these papers, and the AGG checked whether the reasons for exclusion were acceptable. This information is accessible on the website of EFSA (https://food.ec.europa.eu/plants/pesticides/approval-active-substances-safeners-and-synergists/renewal-approval/glyphosate/assessment-group_en).

In summary: From 2010 till June 2020, about 11,300 documents were screened, and the GRG excluded about 10,500. The AGG agreed with most exclusions but asked to add some other papers, either not included in the list or wrongly excluded. About 700 studies remained valid.

- *I agree that this number is out of normal proportion. However, upon checking the exclusions, the reason was frequently "not about glyphosate," "about genetically transformed cultures," "about other pesticides," and alike. This makes the high figure understandable since the selection at the base was not accurate. Papers on other pesticides are probably interesting but constitute indirectly to the final evaluation.*

The second remark is on the overall quality of these documents, both scientific and in the lay press. From an economic point of view, the use of herbicides in agriculture is a financially high-level business in the multiple billionth range. The economic interest of all people involved in the chain of actions is enormous; hence, many people are reluctant to change this economic model. This model contrasts with the need to balance human health and economic interests; this balance is delicate and depends on who is speaking. In this twilight zone, we should always remember the old saying: in case of doubt, Cui Bono or follow the money.

In addition, it takes little conviction that when financial interests are that important, the risk for less ethical and even illegal interventions is substantial; hence, it is not surprising that, in the glyphosate dossier, unsavory practices have occurred. However, the quality of the information on malpractices is also difficult to evaluate and most likely permeated with plainly and deliberately presented wrong data (cf. the email correspondence within Monsanto). In view of the attempts to bring good information—as much as possible—I decided to treat this subject very briefly and make reference only to verifiable information.

In 2019, in a detailed report on the content of the Renewal Assessment Report (RAR) on glyphosate published by Stefan Weber and Helmut Burtscher-Schaden, official instances were openly accused of plagiarism while writing the RAR (https://left.eu/content/uploads/2019/01/Expertise-RAR-Glyphosat-2018-01-11-1.pdf). Parts of RAR were copied and pasted from the documents delivered by the applicants. This accusation caused an overt reaction of the Bundesinstitut für Risikobewertung—BfR (https://www.bfr.bund.de/en/press_information/2019/02/european_assessment_of_glyphosate_is_quality_assured_and_independent___industry_reports_are_routinely_part_of_assessment_reports-239502.html), denying all accusations. This in turn, led to a reaction from members of the EU-PEST-committee (https://mobil.bfr.bund.de/cm/349/open-letter-from-maria-noichl-member-of-the-european-parliament-to-bfr.pdf).

- *Accusations of plagiarism are serious, particularly in sensitive issues like the glyphosate dossier. The analysis showed that 50–80% of the text was copied and pasted from the application documents, which is substantial. It is unlikely that renowned organizations like the BfR would intentionally mismatch documents, and as such, using existing documents in new documents is not illegal provided accurate referencing is given. Apparently, this was not always the case. Taken together, this series of interventions fuels some distrust and uneasiness.*

We know of the Monsanto papers, written by authors paid by Monsanto. The Monsanto papers have been investigated in the USA following lawsuits of victims against Monsanto; this information is publicly available through the action of legal companies, e.g., https://www.wisnerbaum.com/assets/Monsanto%20Roundup%20pages/Secret%20Documents/monsanto-documents-chart-101217.pdf.

Influential papers from EFSA are written by employees of EFSA, other papers were written by scientists engaged in the IARC working group judging the carcinogenicity of glyphosate in Monograph 112 and experts were often selected to defend a particular point of view in US lawsuits thereby obscuring objectivity. Some documents are exemplary. In 2022, an impressive review of the neurological outcome of glyphosate exposure concluded that glyphosate had no effect (Moser et al., 2022a, b). The study was sponsored by the Glyphosate Renewal Group, and one of the authors was a former employee of Monsanto. The authors claim scientific validity, but at least there will remain an atmosphere of distrust. Some of these aspects have been discussed previously (Barnett & Gibson, 2020; Mesnage & Antoniou, 2017), and the need for independent research is stressed (Landrigan & Belpoggi, 2018).

Similarly, different action groups have declared their concerns. The Pesticide Action Network published a press release in 2020 on fraud in German labs during the reapproval of the use of glyphosate in 2017 (https://www.pan-europe.info/sites/pan-europe.info/files/public/resources/press-releases/PR%20with%20LIFE%20logo/20200211%20PR%20Fraud%20in%20LPT%20lab%20Glyphosate_Final.

pdf). Bayer, who took over Monsanto and is the actual main applicant of glyphosate, described their vision of glyphosate (https://www.bayer.com/en/glyphosate/glyphosate-roundup).

On top of this are the lobbying practices within the EU. The actual practice of lobbying has been discussed recently (Hanegraaff & Poletti, 2021; Hanegraaff et al., 2024). A short historical overview of the lobbying practices in the EU was presented to the European Parliament (https://www.europarl.europa.eu/RegData/etudes/etudes/join/2007/393266/IPOL-AFCO_ET(2007)393266_EN.pdf).

Some details mentioned below demonstrate the size and the importance of lobbying.

Lobbying is defined on p5 of the "Green Paper" by the Commission as "all activities carried out with the objective of influencing the policy formulation and decision-making process of the European Institutions" (https://eur-lex.europa.eu/legal-content/EN/TXT/PDF/?uri=CELEX:52006DC0194&qid=1730145884516) and has been discussed in detail recently (https://civitas.org.uk/content/files/Anthony-Chambers-EU-lobbying.pdf). As such, lobbying is, in principle, part of a democratic environment in which individuals, companies, and all possible organizations have the opportunity to express their opinions on matters of common interest in the EU. This initially noble approach became a powerful method used by companies to influence decisions, and it became mainly a method for deregulation or fewer rules, a free market, a higher return on investments, etc. Common societal (health) interests were frequently (if not always) of secondary importance.

This has resulted in some degree of regulation at the EU level, such as the introduction of the Mandatory Transparency Register in July 2021. All members of the Parliament, Commission, and Council must mention all contacts with lobbyists provided that the contact occurred within the boundaries of the buildings of the European organization. Furthermore, the NGO Corporate Europe Observer (CEO—https://corporateeurope.org/en/who-we-are) follows all lobbying activities, and detailed information is available on the website of Lobbyfacts, which is a combined project of CEO and LobbyControl (https://www.lobbyfacts.eu/about). The conclusion is that lobbying has become an integral and established part of European politics.

The return on investment of the lobbying money is difficult to evaluate, although companies will not spend millions of euros for lobbying practices if there is no return value. The European Central Bank compared the data of the Mandatory Register with the AMADEUS databank (a database with financial information on public and private companies across Europe) and could not draw conclusions (https://www.ecb.europa.eu/pub/pdf/scpwps/ecb.wp.2071.en.pdf). Alternatively, William and Macedo concluded that in the USA, the returns on investments are evident, whereas in Europe, they are expected to be (Grotteria, 2023; William & Macedo, 2021).

Some industrial sectors are particularly active in the lobbying business. LobbyFacts.eu published data on this subject in 2022; see Table 9.2.

According to Euronews, the top six lobby spenders in 2023 in Brussels were Meta, €8 million; Apple, €7 million; Bayer, €6 million; Google, €5.5 million; Shell,

Table 9.2 Summary of information on lobbying practices within the EU. The data is from 2022

EU Transparency Register data as of: 19 Sept 2022. A total of 241 results.

#	Name	Head office in	Lobby costs	EP passes	Lobbyists (FTE)	Meetings with EC
1	Bayer AG	Germany	6,500,000€	14	21.7	41
2	Apple Inc.	USA	6,500,000€	9	7.25	77
3	Google	USA	6,000,000€	10	6.45	281
4	Meta Platforms Ireland Limited and its various subsidiaries (f/k/a Facebook Ireland Limited)	Ireland	6,000,000€	8	17.85	176
5	Microsoft Corporation	USA	5,500,000€	4	4.95	174
6	QUALCOMM Incorporated	USA	4,000,000€	5	2.8	50
7	Shell Companies	UK	4,000,000€	4	12	88
8	ExxonMobil Petroleum & Chemical	Belgium	3,500,000€	6	5.5	41
9	Huawei Technologies	China	3,000,000€	11	21	69
10	Volkswagen Aktiengesellschaft	Germany	3,000,000€	5	17.5	78

Top 10 companies declaring highest EU lobby budget in September 2022

LF LobbyFacts.eu

€5.5 million; and Microsoft, €5 million https://www.euronews.com/my-europe/2023/09/11/tech-companies-spend-more-than-100-million-a-year-on-eu-digital-lobbying. Whatever the correct figures are and realizing that the figures are what is officially registered, knowing that there is likely a "less publicly known" gray circuit, a large amount of money is involved. Notably, related to the revenue of the company, Bayer spends 6.5 million EUR, which is less than 0.1% of the 51.05 billion USD turnover in 2024.

Conclusions on the EFSA Documents

It is not easy to obtain a full overview of the large amount of information. The multitude of coding systems, locations, authors, types of documents, abbreviations, etc., hamper easy lectures; hence, material mistakes in this chapter cannot be excluded. Despite this, some general remarks are possible.

For most of these studies, there are few doubts: Glyphosate has little acute toxicity, and it is toxic to aquatic life. For other targets, variable toxicity was found, and in general, there was good agreement with the interpretation found in the RAR and presented in the GRG. Some critical questions, taken up in the text, are feasible and require adequate answers.

As for the 2-year chronic feeding carcinogenicity studies with rats and mice, the experimental information is old or very old. Most of the experiments needed for the actual application date from 1990 to 1995, upfront the first demand for the use of glyphosate as an herbicide. Except for some highly disputed papers, such as the Séralini paper (Séralini et al., 2012) and the ongoing long-term study at the Ramazzinni Institute, nearly no new information is present; hence, the experts have now re-reviewed the information that has been used at least twice before during previous procedures. This has two consequences. (1) Since the judgment in previous procedures by experts and by the EFSA was in favor of (further) the use of glyphosate, it would have been astonishing if any other conclusion had emerged now in the EFSA documents despite the claim that "all evidence is reinvestigated" (2). From the procedures, the ambiguity of the results and conclusions of the experiments were criticized by scientists and lay people; suffice to refer to recent publications on glyphosate evaluation (https://www.generations-futures.fr/wp-content/uploads/2021/11/glyphosate-evaluation-a-severely-skewed-report-v2.pdf). In particular, the reliability of the results of the studies was assessed by Knasmüller and Nersesyan (https://echa.europa.eu/documents/10162/17090/3_glyphosate_eeb_rac60_en.pdf/bff8a83d-5b86-d4ee-ecb3-de873121cfd9?t=1649331309977). The main conclusion of these authors was that only 2 out of 11 studies presented to the AGG and EFSA were reliable.

In the name of human health, industrial partners should invest in better and more recent experiments. There is no answer to the question of why this has not happened.

As demonstrated in Chap. 3, for every scientific paper claiming human toxicity/carcinogenicity, there is a paper denying toxicity, and no distinction in quality is evident; most papers are peer-reviewed and published in respectable journals. It is possibly a sign of exaggerated distrust from my side, but at the end of the day, some suspicion about all glyphosate papers emerges. The situation is somewhat different for animal studies. Most recent studies have been performed by dedicated, well-organized, and expert laboratories in the field, and, according to status and, eventually, obligations within ISO norms, trustworthy results have been obtained. However, the interpretation of the results can be modified according to lines of thinking, which are eventually imposed by the organization that ordered the experiments. As shown in the analysis of the publicly listed animal studies, discrepancies with the RAR are encountered, which means that the animal studies fuel the need for a cautious approach. Taken together, this is the ideal situation in which to apply the precautionary principle.

On the basis of the conclusion of the EFSA, the EU Commission granted a 10-year further use. EFSA made the work for the European Commission easy, stating that there is no human problem. The Commission could, in good thrust, consider the economic debacle as the pivotal argument in the case of a nonrenewal. It is beyond any doubt that members of the Parliament and the Commission are subject to massive lobbying activities, with most pressure coming from the groups in favor of renewal. Taken together, the EFSA has had the courage to contradict all the pressure groups against the further use of glyphosate, but in my humble opinion, the EFSA should have had the courage to decide on precautions. When insufficient data or any other obstacle preventing straightforward reasoning was at hand, the conclusions were always in favor of glyphosate use. Some sentences are exemplary for this approach: "…Suitable data to refine the risk assessment were not available," "….

The experts acknowledge the lack of harmonized methodologies…," "…No international agreed guidelines for the risk assessment of the microbiome are in place in the pesticide area," and "…a higher consumer exposure to residues of glyphosate than the one considered…. It is not EXPECTED that…" The doubt was never interpreted as negative for further use of glyphosate. Admittedly, not one single legal rule was violated, which might have been an important consideration of the experts.

One important item in the advice of EFSA is: "if new information becomes available, the advice will be reviewed." This is praiseworthy but, at the same time, worrisome for a number of reasons. (I) In what field of research does the evidence need to come from that will mark the willingness to review the advice? Epidemiology is probably not accepted since this is considered inadequate. Biology, cell biology, medicine? (II) What is the criterion for accepting new evidence as relevant? (III) MOST IMPORTANT, who will judge that new information is relevant?

Since this is an important item, I asked a question about the reviewing process. My mail and the answer from the DG for health and food safety are given in Chap. 11, Annex 7. In short, the process of re-reviewing is initiated by the Commission upon request of (one of the) member states. Subsequently, the Commission will ask the advice of the member states AND of EFSA on the feasibility of the request. Maybe EFSA, being the direct provider of scientific information to the Commission, does not decide on re-reviewing; its advice will be important. The need for independent research and evaluation has been shown many times (Landrigan & Belpoggi, 2018).

The completeness of the data banks used for the RAR is questionable. In the USA, 35 studies in the field of developmental neurotoxicity were presented to the authorities, of which 9 are NOT presented within the EU (Mie & Ruden, 2023). According to the authors, seven of these nine studies could influence regulatory decisions. Eleven rat and mouse studies were released upon case law (the public files), of which two studies were not present in the data bank of the applicant or the reviewing committee. It cannot be analyzed whether the decision would have been different; at least, there is some doubt about the completeness of the information.

Here again, I asked the question on the completeness of the dossier. My mail and the answer from the Legal Affairs Service are included in Chap. 11, Annex 8.

EFSA's peer review and decision-making have generated many reactions from anti-glyphosate instances, some of which are more emotional than scientific. Among the more fundamental remarks are those from the Health and Environment Alliance (HEAL) from June 2022 (https://www.env-health.org/wp-content/uploads/2022/06/HEAL-How-the-EU-risks-greenlighting-a-pesticide-linked-to-cancer-2022.pdf) and a letter from the Pesticide Action Network (PAN-Europe) to the Executive director of EFSA in July 2023 (https://www.pan-europe.info/sites/pan-europe.info/files/public/resources/Letters/Letter%20to%20B.%20Url%20on%20EFSA%27s%20main%20findings%20on%20glyphosate.pdf).

The value of the HEAL document was apparently sufficient to provoke a response from the ECHA (https://food.ec.europa.eu/document/download/a961296a-e06c-48b8-b955-c32d692ec77c_en?filename=pesticides_renew_glyphosate_letter_20220701_ec-echa_reply.pdf). What follows is a short overview of the points in the HEAL document and the reaction of ECHA. Some reactions are curious.

- HEAL claimed that tumors in carcinogenicity studies were dismissed from the assessment; the ECHA formally denies this claim: No tumors were dismissed.
  - *When the number of tumors was proven to be incorrect, the definition of (the type of) tumor, an error in counting, or a combination of both occurred. It would be surprising that figures from the file reporting the results of the studies were deliberately eliminated. Material mistakes are always possible and this is acceptable and can be corrected in full transparency.*
- The risk assessment committee (RAC) of the ECHA considers the hazardous properties of substances on the basis of the available data and claims that the definition of what constitutes sufficient evidence used by the ECHA is similar to that used by the IARC.
  - *This is consistent with the reason why the ECHA/EFSA and IARC reached different conclusions: The IARC used only public evidence, and the ECHA/EFSA used additional industry studies. The ECHA response states that "the evidence to be weighed differs." It would be good to publish the ultimate approach and the methodology of weighing results.*
- Some paragraphs in the response to HEAL are devoted to the statistical methods.
  - *Avoiding technical descriptions, the main point is whether to use a two-sided test or a one-sided test. In practice, this seems to be an unnecessary problem. A healthy animal exposed to glyphosate can develop cancer, yes or no, which is a digotomic situation. You cannot have something that is less than a cancer; hence, evolution—if there is any—is always in one direction. A protective effect of glyphosate on cancer has not yet been reported. The seriousness of the cancer phenotype may be a continuous variable, but this is different from the yes/no cancer variable.*
- The control group (unexposed animals) in the carcinogenicity studies is "the most important consideration in the testing for increased tumor rates."
  - *This statement is irrefutably true, and each experiment should have a control arm. The reference to historical data is not justified, although it is frequently used.*
- All the data indicate that 2-year-old mice and rats have tumors, which is compatible with all the available evidence that age is an important parameter in tumor growth. Hence, the tumors observed at 2 years of treatment are subject to criticism.
  - *This is undeniable and reflects the relative importance of 2-year carcinogenicity studies. Only with a clear distinction between controls and treatment will a cautious conclusion be accepted. Notably, this is no argument in favor of or against glyphosate. This type of carcinogenicity test has been challenging (Knight* et al., 2006a, b, c; *Suarez-Torres* et al., 2020).

The PAN document highlights the legal background of the EFSA decision and concludes that the EFSA has not followed its own rules. In particular, the use of "a critical area of Concern" or CAoC by EFSA is criticized; the notion of the "precautionary principle" as fixed legally was not respected, the general food law was not respected, and—most importantly—the independence of the EFSA is questioned. To the best of our knowledge, the EFSA has not reacted to this PAN letter.

It is difficult to summarize the above in one statement. The closest comment is that the precautionary principle should have been applied or at least considered. This has not happened.

References

Barnett, J. A., & Gibson, D. L. (2020). Separating the empirical wheat from the pseudoscientific chaff: A critical review of the literature surrounding glyphosate, dysbiosis and wheat-sensitivity. *Frontiers in Microbiology, 11*, 8. https://doi.org/10.3389/fmicb.2020.556729

Bergström, L., Börjesson, E., & Stenström, J. (2011). Laboratory and lysimeter studies of glyphosate and aminomethylphosphonic acid in a sand and a clay soil. *Journal of Environmental Quality, 40*(1), 98–108. https://doi.org/10.2134/jeq2010.0179

Bloem, B. R., & Boonstra, T. A. (2023). The inadequacy of current pesticide regulations for protecting brain health: The case of glyphosate and Parkinson's disease. *Lancet Planetary Health, 7*(12), E948–E949. https://doi.org/10.1016/s2542-5196(23)00255-3

Boksová, A., Kazda, J., Bartoska, J., & Kamler, M. (2023). Effect of glyphosate on the foraging activity of the European honey bee (*Apis mellifera* L.). *Plant, Soil and Environment, 69*(5), 195–201. https://doi.org/10.17221/86/2023-pse

Frederiks, C., & Wesseler, J. H. H. (2019). A comparison of the EU and US regulatory frameworks for the active substance registration of microbial biological control agents. *Pest Management Science, 75*(1), 87–103. https://doi.org/10.1002/ps.5133

Grotteria, M. (2023). Follow the money. *Review of Economic Studies, 40*, 1122. https://doi.org/10.1093/restud/rdad041

Hanegraaff, M., & Poletti, A. (2021). The rise of corporate lobbying in the European Union: An agenda for future research. *JCMS: Journal of Common Market Studies, 59*(4), 839–855. https://doi.org/10.1111/jcms.13132

Hanegraaff, M., Poletti, A., & Van Ommeren, E. (2024). Firms and trade policy lobbying in the European Union. *JCMS: Journal of Common Market Studies, 62*(3), 629–652. https://doi.org/10.1111/jcms.13520

He, X., Tu, Y., Song, Y. W., Yang, G. H., & You, M. D. (2022). The relationship between pesticide exposure during critical neurodevelopment and autism spectrum disorder: A narrative review. *Environmental Research, 203*, 12. https://doi.org/10.1016/j.envres.2021.111902

Hulme, P. E. (2023). Weed resistance to different herbicide modes of action is driven by agricultural intensification. *Field Crops Research, 292*, 10. https://doi.org/10.1016/j.fcr.2023.108819

Klátyik, S., Simon, G., Oláh, M., Mesnage, R., Antoniou, M. N., Zaller, J. G., & Székács, A. (2023). Terrestrial ecotoxicity of glyphosate, its formulations, and co-formulants: Evidence from 2010-2023. *Environmental Sciences Europe, 35*(1), 29. https://doi.org/10.1186/s12302-023-00758-9

Knight, A., Bailey, J., & Balcombe, J. (2006a). Animal carcinogenicity studies: 1. Poor human predictivity. *ATLA: Alternatives to Laboratory Animals, 34*(1), 19–27. https://doi.org/10.1177/026119290603400117

Knight, A., Bailey, J., & Balcombe, J. (2006b). Animal carcinogenicity studies: 2. Obstacles to extrapolation of data to humans. *ATLA: Alternatives to Laboratory Animals, 34*(1), 29–38. https://doi.org/10.1177/026119290603400118

Knight, A., Bailey, J., & Balcombe, J. (2006c). Animal carcinogenicity studies: 3. Alternatives to the bioassay. *ATLA: Alternatives to Laboratory Animals, 34*(1), 39–48. https://doi.org/10.1177/026119290603400119

Kubsad, D., Nilsson, E. E., King, S. E., Sadler-Riggleman, I., Beck, D., & Skinner, M. K. (2019). Assessment of glyphosate induced epigenetic transgenerational inheritance of pathologies and sperm epimutations: Generational toxicology. *Scientific Reports, 9*, 6372. https://doi.org/10.1038/s41598-019-42860.0

La Merrill, M. A., Vandenberg, L. N., Smith, M. T., Goodson, W., Browne, P., Patisaul, H. B., et al. (2020). Consensus on the key characteristics of endocrine-disrupting chemicals as a basis for hazard identification. *Nature Reviews Endocrinology, 16*(1), 45–57. https://doi.org/10.1038/s41574-019-0273-8

Landrigan, P. J., & Belpoggi, F. (2018). The need for independent research on the health effects of glyphosate-based herbicides. *Environmental Health, 17*, 4. https://doi.org/10.1186/s12940-018-0392-z

Madani, N. A., & Carpenter, D. O. (2022). Effects of glyphosate and glyphosate-based herbicides like RoundupTM on the mammalian nervous system: A review. *Environmental Research, 214*, 9. https://doi.org/10.1016/j.envres.2022.113933

Mesnage, R., & Antoniou, M. N. (2017). Facts and fallacies in the debate on glyphosate toxicity. *Frontiers in Public Health, 5*, 7. https://doi.org/10.3389/fpubh.2017.00316

Mesnage, R., Mazzacuva, F., Caldwell, A., Halket, J., & Antoniou, M. N. (2021). Urinary excretion of herbicide co-formulants after oral exposure to roundup MON 52276 in rats. *Environmental Research, 197*, 111103. https://doi.org/10.1016/j.envres.2021.111103

Mie, A., & Ruden, C. (2023). Non-disclosure of developmental neurotoxicity studies obstructs the safety assessment of pesticides in the European Union. *Environmental Health, 22*(1), 21. https://doi.org/10.1186/s12940-023-00994-9

Moser, V., Morris-Schafer, K., Richardson, J., & Li, A. (2022a). Expert panel review of the neurotoxicology of glyphosate in mammalian systems for risk assessment purposes. *Toxicology Letters, 368*, S80–S81. https://doi.org/10.1016/j.toxlet.2022.07.234

Moser, V. C., Morris-Schaffer, K., Richardson, J. R., & Li, A. A. (2022b). Glyphosate and neurological outcomes: A systematic literature review of animal studies. *Journal of Toxicology and Environmental Health, Part B: Critical Reviews, 25*(4), 162–209. https://doi.org/10.1080/10937404.2022.2083739

Motta, E. V. S., Raymann, K., & Moran, N. A. (2018). Glyphosate perturbs the gut microbiota of honey bees. *Proceedings of the National Academy of Sciences of the United States of America, 115*(41), 10305–10310. https://doi.org/10.1073/pnas.1803880115

Motta, E. V. S., Mak, M., De Jong, T. K., Powell, J. E., O'Donnell, A., Suhr, K. J., et al. (2020). Oral or topical exposure to glyphosate in herbicide formulation impacts the gut microbiota and survival rates of honey bees. *Applied and Environmental Microbiology, 86*(18), 21. https://doi.org/10.1128/aem.01150-20

Ofosu, R., Agyemang, E. D., Marton, A., Pasztor, G., Taller, J., & Kazinczi, G. (2023). Herbicide resistance: Managing weeds in a changing world. *Agronomy, 13*(6), 16. https://doi.org/10.3390/agronomy13061595

Peillex, C., & Pelletier, M. (2020). The impact and toxicity of glyphosate and glyphosate-based herbicides on health and immunity. *Journal of Immunotoxicology, 17*(1), 163–174. https://doi.org/10.1080/1547691x.2020.1804492

Robinson, C., Portier, C. J., Cavoski, A., Mesnage, R., Roger, A., Clausing, P., et al. (2020). Achieving a high level of protection from pesticides in Europe: Problems with the current risk assessment procedure and solutions. *European Journal of Risk Regulation, 11*(3), 450–480. https://doi.org/10.1017/err.2020.18

Samsel, A., & Seneff, S. (2013). Glyphosate's suppression of cytochrome P450 enzymes and amino acid biosynthesis by the gut microbiome: Pathways to modern diseases. *Entropy, 15*(4), 1416–1463. https://doi.org/10.3390/e15041416

Schweizer, M., Brilisauer, K., Triebskorn, R., Forchhammer, K., & Kohler, H. R. (2019). How glyphosate and its associated acidity affect early development in zebrafish (Danio rerio). *PeerJ, 7*, 25. https://doi.org/10.7717/peerj.7094

Schwientek, M., Rügner, H., Haderlein, S. B., Schulz, W., Wimmer, B., Engelbart, L., et al. (2024). Glyphosate contamination in European rivers not from herbicide application? *Water Research, 263*, 10. https://doi.org/10.1016/j.watres.2024.122140

Séralini, G. E., Clair, E., Mesnage, R., Gress, S., Defarge, N., Malatesta, M., et al. (2012). RETRACTED: Long term toxicity of a Roundup herbicide and a Roundup-tolerant genetically modified maize (Retracted Article). *Food and Chemical Toxicology, 50*(11), 4221–4231. https://doi.org/10.1016/j.fct.2012.08.005

Straw, E. A., Carpentier, E. N., & Brown, M. J. F. (2021). Roundup causes high levels of mortality following contact exposure in bumble bees. *Journal of Applied Ecology, 58*(6), 1167–1176. https://doi.org/10.1111/1365-2664.13867

Straw, E. A., Mesnage, R., Brown, M. J. F., & Antoniou, M. N. (2023). No impacts of glyphosate or *Crithidia bombi*, or their combination, on the bumblebee microbiome. *Scientific Reports, 13*(1), 8. https://doi.org/10.1038/s41598-023-35304-3

Suarez-Torres, J. D., Jimenez-Orozco, F. A., & Ciangherotti, C. E. (2020). The 2-year rodent bioassay in drug and chemical carcinogenesis testing: Sensitivity, according to the framework of carcinogenic action. *Toxicology Mechanisms and Methods, 30*(6), 462–475. https://doi.org/10.1080/15376516.2020.1760986

William, A. W., & Macedo, F. S. (2021). Does it pay to lobby? Examining the link between firm lobbying and firm profitability in the European Union. *Journal of European Public Policy, 28*(12), 1993–2010. https://doi.org/10.1080/13501763.2020.1824012

Chapter 10
Future of Glyphosate

Some fundamental questions remain: (1) What is the actual situation of glyphosate use in the EU and worldwide? (2) Are there any alternatives to glyphosate? (3) What is the weight of the results of actual research in the evaluation process of glyphosate at the EU level? (4) Under the assumption that glyphosate is banned from the market, what are the consequences? Can we ban glyphosate? Do we have a valid alternative?

Prediction on the Use of Glyphosate, Role of Resistance

Only 2 years after glyphosate-resistant crops were first used in 1996, Powles et al. published on the resistance in rigid ryegrass to glyphosate (Powles et al., 1998). Currently, more than 40 examples of the resistance of herbs to glyphosate have been described, and several molecular mechanisms of resistance have been described in detail (Beckie et al., 2019; Duke & Powles, 2008; Heap & Duke, 2018; Hertel et al., 2021; Hulme, 2023; Ofosu et al., 2023; Peterson et al., 2018; Powles, 2010). The situation is visualized by Heap and Duke (see Fig. 10.1). The occurrence of resistance of weeds against herbicides is registered in the International Herbicide-Resistant Weed Database (https://www.weedscience.org/Pages/ReferenceView.aspx). As O'Duke noted, "Darwinian reality spoils a good thing" (Duke, 2018), referring to the genetic adaptability and resistance of herbs to glyphosate, thereby reducing its usability in the field.

The occurrence of resistance against herbicides is not a property of glyphosate alone: Every herbicide induces resistance. Owing to its abundant use since the 1990s, resistance to glyphosate is more important or, at least, more discussed. The consequences for further use are proportionally great.

This resistance situation with glyphosate is in some way comparable to the situation with antibiotics (Baran et al., 2023). When the first antibiotic was introduced

N. Fraeyman, *Glyphosate 2023–2033*,
https://doi.org/10.1007/978-3-031-97540-0_10

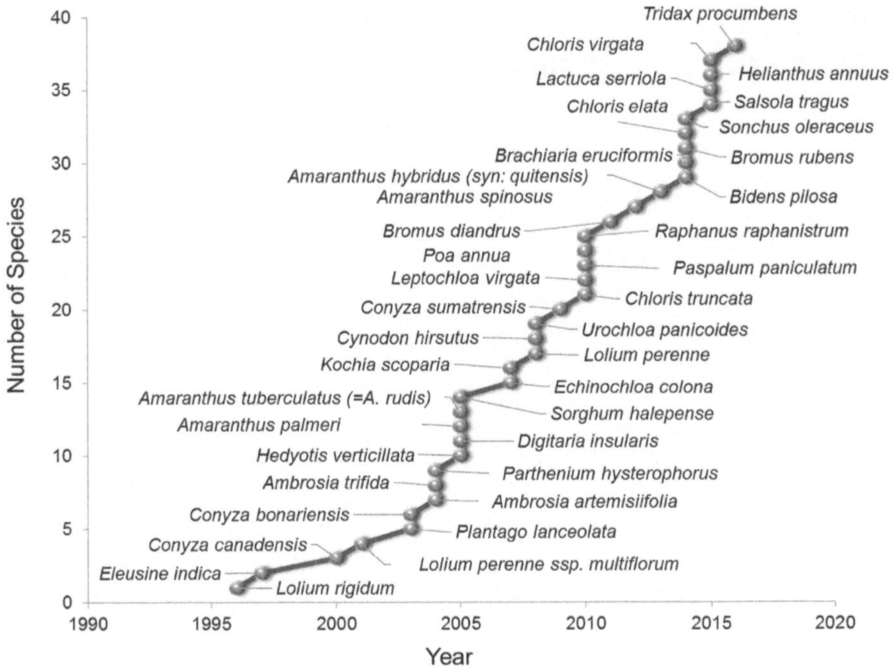

Fig. 10.1 Time-dependent evolution of the resistance of plants to glyphosate (Taken from Heap and Duke (2018)

on the market (penicillin in 1928), success was enormous, and as a consequence, antibiotics were used abundantly. In his Nobel lecture in 1945, Fleming himself warned against overuse and the rise of resistance. Currently, bacterial resistance is a global threat, and the WHO has declared bacterial resistance a near-catastrophic situation in health care. Antibiotic overuse creates pressure on the genetics of microbes and reflects continuous random genetic modifications followed by Darwinian selection for resistance. The flexibility of unicellular microorganisms is further illustrated by the fact that cross-resistance against different types of antibiotics with different molecular targets occurs, further strengthening this problem. The solution is either to reduce the use of antibiotics and reduce the genetic pressure and subsequently reduce resistance or to search for other antibiotics. This "de-resistance" process might take at least as long as the time leading to resistance, i.e., probably decades.

Something very comparable is happening with glyphosate. The appearance of resistant weeds after the introduction of glyphosate demonstrates the genetic flexibility of the plants and jeopardizes the use of glyphosate. The solutions for further controlling herbs on short notice are limited: Either the doses of glyphosate (unit of weight/unit of land surface) are increased or glyphosate is combined with other pesticides; either glyphosate is abandoned, and other pesticides are used. The solution in the longer term is the reduction of the doses in combination with other methods to remove weeds. No solution is optimal and might ultimately lead to greater problems than what we currently have with glyphosate alone. The need for a safe chemical alternative to glyphosate is urgent, comparable to the search for other antibiotics.

Alternatives to Glyphosate

Abbas et al. (2018) described the different possibilities for herb control, including nonchemical methods such as manual weeding and hoeing, the use of chemical or biological alternatives, and the optimal combination of herbicides and other chemicals. According to some models, including the Australian model, agriculture without glyphosate is possible (Beckie et al., 2020, 2021).

Manual and mechanical weeding are preferred from an ecological point of view. However, energy consumption when mechanical weeding is used should be considered, while the costs for consumers will increase significantly. Additionally, this type of weed control is only possible when crops are grown in rows or as separate entities, such as fruit trees or in a vineyard (Jacquet et al., 2021). With cultures such as great planes with thousands of acres of plants, chemical treatment is indicated. The question of alternatives for glyphosate has been documented extensively (Fogliatto et al., 2020).

In view of this growing resistance, the search for chemical alternatives will continue and probably increase (Berestetskiy, 2023), but this has not yet resulted in the identification of miracle molecules comparable to glyphosate. Phillips McDougall showed in 2016 that the costs and time needed to bring to the market a new pesticide is comparable to marketing a new pharmaceutical and is about 10 years (https://www.seedquest.com/News/pdf/2016/PMcD.pdf). In view of the success of glyphosate, the sense of urgency was (possibly) absent in the last decade. Some possibilities have been proposed, although with limited success, e.g., haloacetates (Turek et al., 2018), some oil-based pesticides (Maes et al., 2021), and protoporphyrinogen (PPO) inhibitors (Barker et al., 2023). Currently, a commercially available product, named Roundup, is currently available, but it does not contain GLY but pelargonic (nonanoic) acid.

The EU promotes the use of "low-risk compounds" (Damalas & Koutroubas, 2018; Marchand, 2015; Robin & Marchand, 2019, 2022, 2023; Romanazzi et al., 2022), and a list of possible candidates has been published (https://eur-lex.europa.eu/legal-content/EN/TXT/?uri=CELEX:52018XC0727(01)). Among the low-risk compounds are biological pesticides promoted as ecologically acceptable alternatives to chemical pesticides (Seiber et al., 2014), and the EU has adopted new rules for facilitating the approval of these compounds (https://food.ec.europa.eu/news/eu-commission-adopts-new-rules-facilitate-approval-biological-pesticides-2022-08-31_en). The EU will furthermore invest substantially in research for finding new biopesticides by 2030 (https://croplifeeurope.eu/wp-content/uploads/2020/11/CropLifeEurope-2030-Commitments.pdf) and by funding

European projects such as BIOBESTcide, which gathers ten partners (https://www.biobesticide.eu/partners/).

Allelopathy, i.e., making use of the antagonistic effect on the growth of weeds by coculture with suitable plants, is a promising approach (Ain et al., 2023; Araniti et al., 2012; Cheng & Cheng, 2015; Khanh et al., 2007; Kostina-Bednarz et al., 2023; Li et al., 2017; Pierik et al., 2013; Quiroz et al., 2019; Sahrir et al., 2023), but the search is still ongoing (Hickman et al., 2021). To date, applications are possible and promising, but the same high-quality properties as those of glyphosate have not yet been reached.

The need for alternatives and/or other methods to tackle the "glyphosate problem" is taken up by many authors and reformulated as "integrated weed management" or "integrative weed control" which means that not one solution to replace glyphosate will be sufficient, but multiple applications according to specific field conditions will be needed (Swanton et al., 2008; Westwood et al., 2018) as described recently (Ofosu et al., 2023). Integrative weed control should be considered as a tiered process with (1) prevention (prevent future growth of herbs by removing seeds of flowering herbs, although this will jeopardize the honey bee), (2) detailed monitoring of the fields to restrict application of the herbicide only at the location where the herbs grow, and (3) a combination of mechanical and biological/biotechnological methods (pelargonic acid, vinegar, flame, steam, or hot foam weeding). Ultimately, if this fails, the use of herbicides and glyphosate is the herbicide of choice while reducing the exposure of non-target sites and humans. (https://content.ces.ncsu.edu/are-there-alternatives-to-glyphosate-for-weed-control-in-landscapes).

Despite the attractiveness of the term "integrated weed management," it is not very clear what this term means exactly in practice. At variance, it is called biological agriculture, although this theme is filled in at random, and many applications are available. Lake and Minteer investigated this point over a period of 30 years and concluded somewhat pessimistically: "*Despite numerous examples of successful programs … there was no increase in the number of studies reported annually over time*" (Lake & Minteer, 2018). It is likely to be (part of) the solution to the glyphosate problem, although, at this moment, much fundamental and application research has to be done. The conclusion is that glyphosate cannot be replaced at this moment.

Doing Nothing Is Not a Good Idea; Banning Glyphosate Tomorrow Is Worse

If glyphosate were removed tomorrow and bearing the actual state of the art in agriculture in mind, significant economic losses for farmers and for the whole society would follow. This has been investigated via a range of economic models (Brookes et al., 2017; Jacquet et al., 2021; Walsh & Kingwell, 2021; Ye et al., 2021), although the outcome of the calculations has been questioned (Finger et al., 2023). Oxford Economics calculated that the impact of banning glyphosate in the UK would lead to a significant reduction in the production of cereals (−15%) and wheat (−12%) and that the overall output of the UK's farmers would be reduced by 3.8% (https://oeservices.oxfordeconomics.com/publication/open/276397).

Abandoning glyphosate in Europe now would have detrimental consequences (Wynn & Webb, 2022). The price of consumable vegetables is expected to increase significantly (up to 20%), and, in addition, overall toxicity would even further increase because of the necessary use of other herbicides with a much higher toxicity profile than glyphosate. However, Böcker et al. assessed the glyphosate ban on maize and wheat production in Switzerland and concluded that the economic consequences are rather limited (Böcker et al., 2018, 2019, 2020; Finger et al., 2023). The ban and re-introduction of glyphosate after 1 year in Sri Lanka is a good illustration of what can happen when banning glyphosate (Dorlach & Gunasekara, 2023).

The use of glyphosate in Europe and in daily practice includes applications before crops are sown and in less accessible zones, such as railway tracks. In some parts of Europe, glyphosate is applied at the end of the culture season and shortly before harvest to synchronize the growth process and speed up the drying process after harvest but this is no longer permitted (https://food.ec.europa.eu/plants/pesticides/approval-active-substances-safeners-and-synergists/renewal-approval/glyphosate_en#:~:text=may%20be%20used%20in%20non,not%20permitted%20in%20the%20EU). The use of glyphosate selectively, e.g., for invasive species or for localized herb growth, has been proposed (Pergl et al., 2020).

The restrictions on the further use of glyphosate are mentioned in the EU report and include:

1. The *prohibition for use as a desiccant* (with the intention to control the time point of harvest or to optimize the threshing).
2. The *setting of maximum limits for five impurities in glyphosate* (i.e., in the material manufactured). These limits are set to ensure that humans and the environment are fully protected.
3. The *requirement for Member States to pay particular attention to specific aspects when carrying out risk assessments* (e.g., the protection of small herbivorous mammals, such as voles, and non-target plants, such as wildflowers) *and to set risk mitigation measures* to ensure that non-target organisms and the environment are protected.
4. The *setting of maximum application rates* that shall not be exceeded unless the outcome of the risk assessment undertaken for the specific uses for which authorization is applied demonstrates that a higher rate does not lead to any unacceptable effects on small herbivorous mammals.
5. An *obligation for the applicant to submit information on the possible indirect impacts on biodiversity* within 3 years from the availability of an appropriate guidance document. The Commission will request EFSA to develop the necessary guidance.

On the keep-glyphosate site, several interesting suggestions have been made. Meftaul et al. and Parven et al. argued that further use of glyphosate is acceptable if the adjuvant is changed from polyoxyethylene tallow amine (conveniently abbreviated as tallow amine or POEA) to a less toxic molecule, assuming that the toxicity of glyphosate is increased by the adjuvant (Meftaul et al., 2020; Parven et al., 2024). Pergl argued in their commentary that glyphosate is useful for specific applications, such as the battle against aggressive alien species or when manual/mechanical removal is not feasible (Pergl et al., 2020) (https://neobiota.pensoft.net/article/51823/).

The banning of glyphosate is not comparable to the banning of atrazine 20 years ago (Mahé et al., 2020). An initial reduction in harvests was found, which was compensated very quickly by replacing atrazine with glyphosate: One chemical was replaced by another. However, to the best of our knowledge, no acceptable chemical alternative for glyphosate is currently available. Hence, the economic damage would be much greater.

Currently, feeding the human population is a challenge, and it is taken for granted that this cannot be responded to adequately without glyphosate (Schneider et al., 2023). Given the rising Mondial population, support for the further use of glyphosate will even increase. The option to change feeding and eating habits from predominantly carnivorous to more herbivorous is favorable, as suggested by the EAT-Lancet Commission, which will allow us to feed people in developing countries and, at the same time, increase our health (Libera et al., 2021; Tilman & Clark, 2015; Willett et al., 2019). In view of the present status of modern society, this seems unrealizable even in the mid- to long-term, and, moreover, the link between the consumption of red meat and cancer has been questioned (Leroy & Cofnas, 2020). In theory, a change in eating habits would be beneficial not only for human health but also for reducing the need for cattle feed, which constitutes approximately 80% of actual agricultural activities. Regardless, the need for an alternative to glyphosate is necessary.

There are many arguments both in favor of and against the use of glyphosate; hence, making a point of view will always lead to controversy, and there is no escape route. The only way is to replace glyphosate with methods/other chemicals with similar or comparable beneficial properties but without human toxicological properties. If no action is taken now (whatever that may mean), the toxicological effects will remain, especially for farmers and their families and all those closely involved in the glyphosate business (production, packaging, transport, spraying, and harvesting). It is easy to predict that in 2030, i.e., at the time the dossier for prolongation of using glyphosate will be submitted, aiming at the prolongation beyond December 2033, many of the discussions in this book will be recycled.

All this ends up in a delicate balance. On the one hand, anthropogenic activities bring prosperity, better health, longevity, etc., while on the other hand, some health benefits are sacrificed, likely not in the short term but possibly in the long term. In this respect, it is the multitude of chemicals that are present in the environment that is to blame, among which glyphosate is a prominent member.

References

Abbas, T., Zahir, Z. A., Naveed, M., & Kremer, R. J. (2018). Limitations of existing weed control practices necessitate development of alternative techniques based on biological approaches. In *Advances in agronomy* (Vol. 147, pp. 239–280). Academic Press.

Ain, Q., Mushtaq, W., Shadab, M., & Siddiqui, M. B. (2023). Allelopathy: An alternative tool for sustainable agriculture. *Physiology and Molecular Biology of Plants, 17*, 495. https://doi.org/10.1007/s12298-023-01305-9

Araniti, F., Sorgona, A., Lupini, A., & Abenavoli, M. R. (2012). Screening of Mediterranean wild plant species for allelopathic activity and their use as bio-herbicides. *Allelopathy Journal, 29*(1), 107–123.

Baran, A., Kwiatkowska, A., & Potocki, L. (2023). Antibiotics and bacterial resistance—A short story of an endless arms race. *International Journal of Molecular Sciences, 24*(6), 34. https://doi.org/10.3390/ijms24065777

Barker, A. L., Pawlak, J., Duke, S. O., Beffa, R., Tranel, P. J., Wuerffel, J., et al. (2023). Discovery, mode of action, resistance mechanisms, and plan of action for sustainable use of Group 14 herbicides. *Weed Science, 16*, 173. https://doi.org/10.1017/wsc.2023.15

Beckie, H. J., Ashworth, M. B., & Flower, K. C. (2019). Herbicide resistance management: Recent developments and trends. *Plants, 8*(6), 161. https://doi.org/10.3390/plants8060161

Beckie, H. J., Flower, K. C., & Ashworth, M. B. (2020). Farming without glyphosate? *Plants, 9*(1), 96. https://doi.org/10.3390/plants9010096

Beckie, H. J., Ashworth, M. B., & Flower, K. C. (2021). The global challenge of field crop production with limited herbicides: An Australian perspective. *Weed Research, 61*(2), 88–91. https://doi.org/10.1111/wre.12421

Berestetskiy, A. (2023). Modern approaches for the development of new herbicides based on natural compounds. *Plants, 12*(2), 24. https://doi.org/10.3390/plants12020234

Böcker, T., Britz, W., & Finger, R. (2018). Modelling the effects of a glyphosate ban on weed management in silage maize production. *Ecological Economics, 145*, 182–193. https://doi.org/10.1016/j.ecolecon.2017.08.027

Böcker, T., Möhring, N., & Finger, R. (2019). Herbicide free agriculture? A bio-economic modelling application to Swiss wheat production. *Agricultural Systems, 173*, 378–392. https://doi.org/10.1016/j.agsy.2019.03.001

Böcker, T., Britz, W., Möhring, N., & Finger, R. (2020). An economic and environmental assessment of a glyphosate ban for the example of maize production. *European Review of Agricultural Economics, 47*(2), 371–402. https://doi.org/10.1093/erae/jby050

Brookes, G., Taheripour, F., & Tyner, W. E. (2017). The contribution of glyphosate to agriculture and potential impact of restrictions on use at the global level. *GM Crops & Food: Biotechnology in Agriculture and the Food Chain, 8*(4), 216–228. https://doi.org/10.1080/21645698.2017.1390637

Cheng, F., & Cheng, Z. H. (2015). Research progress on the use of plant allelopathy in agriculture and the physiological and ecological mechanisms of allelopathy. *Frontiers in Plant Science, 6*, 16. https://doi.org/10.3389/fpls.2015.01020

Damalas, C. A., & Koutroubas, S. D. (2018). Current status and recent developments in biopesticide use. *Agriculture, 8*(1), 6. https://doi.org/10.3390/agriculture8010013

Dorlach, T., & Gunasekara, S. (2023). The politics of glyphosate regulation: Lessons from Sri Lanka's short-lived ban. *Globalization and Health, 19*(1), 14. https://doi.org/10.1186/s12992-023-00981-2

Duke, S. O. (2018). The history and current status of glyphosate. *Pest Management Science, 74*(5), 1027–1034. https://doi.org/10.1002/ps.4652

Duke, S. O., & Powles, S. B. (2008). Glyphosate-resistant weeds and crops. *Pest Management Science, 64*(4), 317–318. https://doi.org/10.1002/ps.1561

Finger, R., Möhring, N., & Kudsk, P. (2023). Glyphosate ban will have economic impacts on European agriculture but effects are heterogenous and uncertain. *Communications Earth & Environment, 4*(1), 9. https://doi.org/10.1038/s43247-023-00951-x

Fogliatto, S., Ferrero, A., & Vidotto, F. (2020). Current and future scenarios of glyphosate use in Europe: Are there alternatives? In D. L. Sparks (Ed.), *Advances in agronomy* (Vol. 163, pp. 219–278). Academic Press Ltd/Elsevier Science Ltd.

Heap, I., & Duke, S. O. (2018). Overview of glyphosate-resistant weeds worldwide. *Pest Management Science, 74*(5), 1040–1049. https://doi.org/10.1002/ps.4760

Hertel, R., Gibhardt, J., Martienssen, M., Kuhn, R., & Commichau, F. M. (2021). Molecular mechanisms underlying glyphosate resistance in bacteria. *Environmental Microbiology, 23*(6), 2891–2905. https://doi.org/10.1111/1462-2920.15534

Hickman, D. T., Rasmussen, A., Ritz, K., Birkett, M. A., & Neve, P. (2021). Review: Allelochemicals as multi-kingdom plant defence compounds: Towards an integrated approach. *Pest Management Science, 77*(3), 1121–1131. https://doi.org/10.1002/ps.6076

Hulme, P. E. (2023). Weed resistance to different herbicide modes of action is driven by agricultural intensification. *Field Crops Research, 292*, 10. https://doi.org/10.1016/j.fcr.2023.108819

Jacquet, F., Delame, N., Vita, J. L., Huyghe, C., & Reboud, X. (2021). The micro-economic impacts of a ban on glyphosate and its replacement with mechanical weeding in French vineyards. *Crop Protection, 150*, 9. https://doi.org/10.1016/j.cropro.2021.105778

Khanh, T. D., Elzaawely, A. A., Chung, I. M., Ahn, J. K., Tawata, S., & Xuan, T. D. (2007). Role of allelochemicals for weed management in rice. *Allelopathy Journal, 19*(1), 85–95.

Kostina-Bednarz, M., Plonka, J., & Barchanska, H. (2023). Allelopathy as a source of bioherbicides: Challenges and prospects for sustainable agriculture. *Reviews in Environmental Science and Bio/Technology, 22*(2), 471–504. https://doi.org/10.1007/s11157-023-09656-1

Lake, E. C., & Minteer, C. R. (2018). A review of the integration of classical biological control with other techniques to manage invasive weeds in natural areas and rangelands. *Biocontrol, 63*(1), 71–86. https://doi.org/10.1007/s10526-017-9853-5

Leroy, F., & Cofnas, N. (2020). Should dietary guidelines recommend low red meat intake? *Critical Reviews in Food Science and Nutrition, 60*(16), 2763–2772. https://doi.org/10.1080/10408398.2019.1657063

Li, W., He, S. Q., Cheng, X. P., & Song, W. H. (2017). Role of allelopathy in plant invasions in natural conditions. *Allelopathy Journal, 41*(2), 249–258. https://doi.org/10.26651/2017-41-2-1100

Libera, J., Ilowiecka, K., & Stasiak, D. (2021). Consumption of processed red meat and its impact on human health: A review. *International Journal of Food Science and Technology, 56*(12), 6115–6123. https://doi.org/10.1111/ijfs.15270

Maes, C., Meersmans, J., Lins, L., Bouquillon, S., & Fauconnier, M. L. (2021). Essential oil-based bioherbicides: Human health risks analysis. *International Journal of Molecular Sciences, 22*(17), 12. https://doi.org/10.3390/ijms22179396

Mahé, I., Gauvrit, C., Angevin, F., & Chauvel, B. (2020). What lessons can be learned from the atrazine ban in the context of the future withdrawal of glyphosate? *Cahiers Agricultures, 29*, 9. https://doi.org/10.1051/cagri/2020026

Marchand, P. A. (2015). Basic substances: An opportunity for approval of low-concern substances under EU pesticide regulation. *Pest Management Science, 71*(9), 1197–1200. https://doi.org/10.1002/ps.3997

Meftaul, I. M., Venkateswarlu, K., Dharmarajan, R., Annamalai, P., Asaduzzaman, M., Parven, A., & Megharaj, M. (2020). Controversies over human health and ecological impacts of glyphosate: Is it to be banned in modern agriculture? *Environmental Pollution, 263*, 18. https://doi.org/10.1016/j.envpol.2020.114372

Ofosu, R., Agyemang, E. D., Marton, A., Pasztor, G., Taller, J., & Kazinczi, G. (2023). Herbicide resistance: Managing weeds in a changing world. *Agronomy, 13*(6), 16. https://doi.org/10.3390/agronomy13061595

Parven, A., Meftaul, I. M., Venkateswarlu, K., & Megharaj, M. (2024). Herbicides in modern sustainable agriculture: Environmental fate, ecological implications, and human health concerns. *International journal of Environmental Science and Technology, 22*, 1181. https://doi.org/10.1007/s13762-024-05818-y

Pergl, J., Hartel, H., Pysek, P., & Stejskal, R. (2020). Don't throw the baby out with the bathwater – Ban of glyphosate use depends on context. *NeoBiota, 56*, 27–29. https://doi.org/10.3897/neobiota.56.51823

Peterson, M. A., Collavo, A., Ovejero, R., Shivrain, V., & Walsh, M. J. (2018). The challenge of herbicide resistance around the world: A current summary. *Pest Management Science, 74*(10), 2246–2259. https://doi.org/10.1002/ps.4821

Pierik, R., Mommer, L., & Voesenek, L. (2013). Molecular mechanisms of plant competition: Neighbour detection and response strategies. *Functional Ecology, 27*(4), 841–853. https://doi.org/10.1111/1365-2435.12010

Powles, S. B. (2010). Gene amplification delivers glyphosate-resistant weed evolution. *Proceedings of the National Academy of Sciences of the United States of America, 107*(3), 955–956. https://doi.org/10.1073/pnas.0913433107

Powles, S. B., Lorraine-Colwill, D. F., Dellow, J. J., & Preston, C. (1998). Evolved resistance to glyphosate in rigid ryegrass (Lolium rigidum) in Australia. *Weed Science, 46*(5), 604–607. https://doi.org/10.1017/s0043174500091165

Quiroz, R. D., Maldonado, J. J. C., Alanis, M. D. R., Torres, J. A., & Saldivar, R. P. (2019). Fungi-based biopesticides: Shelf-life preservation technologies used in commercial products. *Journal of Pest Science, 92*(3), 1003–1015. https://doi.org/10.1007/s10340-019-01117-5

Robin, D. C., & Marchand, P. A. (2019). Evolution of the biocontrol active substances in the framework of the European Pesticide Regulation (EC) No. 1107/2009. *Pest Management Science, 75*(4), 950–958. https://doi.org/10.1002/ps.5199

Robin, D. C., & Marchand, P. A. (2022). Expansion of the low-risk substances in the framework of the European Pesticide Regulation (EC) No. 1107/2009. *European Journal of Risk Regulation, 13*(3), 514–531. https://doi.org/10.1017/err.2021.58

Robin, D. C., & Marchand, P. A. (2023). The slow decrease of active substance candidates for substitution in the framework of the European Pesticide Regulation (EC) No. 1107/2009. *European Journal of Risk Regulation, 22*, 191. https://doi.org/10.1017/err.2021.20

Romanazzi, G., Orconneau, Y., Moumni, M., Davillerd, Y., & Marchand, P. A. (2022). Basic substances, a sustainable tool to complement and eventually replace synthetic pesticides in the management of pre and postharvest diseases: Reviewed instructions for users. *Molecules, 27*(11), 39. https://doi.org/10.3390/molecules27113484

Sahrir, M. A. S., Yusoff, N., & Azizan, K. A. (2023). Allelopathy activity under laboratory, greenhouse and field conditions: A review. *AIMS Agriculture and Food, 8*(1), 78–104. https://doi.org/10.3934/agrfood.2023004

Schneider, K., Barreiro-Hurle, J., & Rodriguez-Cerezo, E. (2023). Pesticide reduction amidst food and feed security concerns in Europe. *Nature Food, 4*(9), 746–750. https://doi.org/10.1038/s43016-023-00834-6

Seiber, J. N., Coats, J., Duke, S. O., & Gross, A. D. (2014). Biopesticides: State of the art and future opportunities. *Journal of Agricultural and Food Chemistry, 62*(48), 11613–11619. https://doi.org/10.1021/jf504252n

Swanton, C. J., Mahoney, K. J., Chandler, K., & Gulden, R. H. (2008). Integrated weed management: Knowledge-based weed management systems. *Weed Science, 56*(1), 168–172. https://doi.org/10.1614/ws-07-126.1

Tilman, D., & Clark, M. (2015). Food, agriculture & the environment: Can we feed the world & save the earth? *Daedalus, 144*(4), 8–23. https://doi.org/10.1162/DAED_a_00350

Turek, M., Biczak, R., Pawlowska, B., Rozycka-Sokolowska, E., Marciniak, B., Deska, M., et al. (2018). Ammonium haloacetates – An alternative to glyphosate? *Chemosphere, 194*, 650–656. https://doi.org/10.1016/j.chemosphere.2017.12.055

Walsh, A., & Kingwell, R. (2021). Economic implications of the loss of glyphosate and paraquat on Australian mixed enterprise farms. *Agricultural Systems, 193*, 9. https://doi.org/10.1016/j.agsy.2021.103207

Westwood, J. H., Charudattan, R., Duke, S. O., Fennimore, S. A., Marrone, P., Slaughter, D. C., et al. (2018). Weed management in 2050: Perspectives on the future of weed science. *Weed Science, 66*(3), 275–285. https://doi.org/10.1017/wsc.2017.78

Willett, W., Rockstrom, J., Loken, B., Springmann, M., Lang, T., Vermeulen, S., et al. (2019). Food in the Anthropocene: The EAT-Lancet Commission on healthy diets from sustainable food systems. *Lancet, 393*(10170), 447–492. https://doi.org/10.1016/s0140-6736(18)31788-4

Wynn, S., & Webb, E. (2022). Impact assessment of the loss of glyphosate within the EU: A literature review. *Environmental Sciences Europe, 34*(1), 91. https://doi.org/10.1186/s12302-022-00667-3

Ye, Z. W., Wu, F., & Hennessy, D. A. (2021). Environmental and economic concerns surrounding restrictions on glyphosate use in corn. *Proceedings of the National Academy of Sciences of the United States of America, 118*(18), 9. https://doi.org/10.1073/pnas.2017470118

Chapter 11
List of Annexes 1–8

Annex 1: Abbreviations

The list of abbreviations is not exhaustive. Many organizations and/or regulatory instances use in-house acronyms and definitions. The applications of these abbreviations can be found on dedicated websites.

AAOEL	Acute acceptable operation exposure level
ADI	Allowable/acceptable daily intake
ALARA	As low as reasonably achievable
ALARP	As low as reasonably possible
AOEL	Acceptable operator exposure level
ARfD	Acute reference dose
BMDL	Benchmark dose limit
BMR	Benchmark response
DNEL	Derived minimal affect level
EDI	Estimated daily intake
EQS	Environmental quality standards
ETO	Ecological threshold option
HEC	Human equivalent concentration
HED	Human equivalent dose
IEDI	Internal estimated daily intake
LD50	Lethal dose for 50% of the animals
LEL	Lowest effect level
LOAEL	Lowest observed adverse effect level
LOEL	Lowest observed effect level
LOQ	Limit of quantification
MABEL	Minimum anticipated biological effect level
MCL	Maximum contaminant level

(continued)

N. Fraeyman, *Glyphosate 2023–2033*,
https://doi.org/10.1007/978-3-031-97540-0_11

MOA	Mode of action
MOE	Margin of exposure
MoS	Margin of safety
MRL	Maximum residue limits
MTD	Maximum tolerated dose
NOAEL	No observed adverse effect level
NOEC	No observed effect concentration
NOEL	No observed effect level
PEC	Predicted environmental concentration (in air, soil, sediment, etc.)
PMTDI	Provisional maximum tolerable daily intake
PNEC	Predicted noneffect concentrations
POD	Point of departure
PTWI	Provisional tolerable weekly intake
RAC	Regulatory acceptable concentration
RfD	Reference dose
TDI	Tolerable daily intake
TMDL	Total maximum daily load

Annex 2: Legal Obligations Related to the Glyphosate Dossier

The following Directives and Regulations are important for renewing the approval of the use of glyphosate. A distinction is made between Regulations (binding and applicable in all member states without the need for national legislation) and Directives (legal instruments that define goals to be achieved by the member states within a certain time window; member states can choose how to implement the Directive within the deadline).

- Regulation (EU) 2023/2660 (https://eur-lex.europa.eu/legal-content/EN/TXT/?uri=CELEX%3A32023R2660).
  Commission Implementing Regulation (EU) 2023/2660 of 28 November 2023 renewing the approval of the active substance glyphosate in accordance with Regulation (EC) No. 1107/2009 of the European Parliament and of the Council and amending Commission Implementing Regulation (EU) No 540/2011
- Regulation (EC) 1107/2009 (https://eur-lex.europa.eu/eli/reg/2009/1107/oj/eng).
  Regulation (EC) No 1107/2009 of the European Parliament and of the Council of 21 October 2009 concerning the placing of plant protection products on the market and repealing Council Directives 79/117/EEC and 91/414/EEC
- Regulation (EU) 2017/2324 (https://eur-lex.europa.eu/eli/reg_impl/2017/2324/oj/eng).

Commission Implementing Regulation (EU) 2017/2324 of 12 December 2017 renewing the approval of the active substance glyphosate in accordance with Regulation (EC) No 1107/2009 of the European Parliament and of the Council concerning the placing of plant protection products on the market, and amending the Annex to Commission Implementing Regulation (EU) No 540/2011.

- Regulation (EU) 540/2011 (https://eur-lex.europa.eu/eli/reg_impl/2011/540/oj/eng).
  Commission Implementing Regulation (EU) No 540/2011 of 25 May 2011 implementing Regulation (EC) No 1107/2009 of the European Parliament and of the Council as regards the list of approved active substances.
- Regulation (EU) 844/2012 (https://eur-lex.europa.eu/eli/reg_impl/2012/844/oj/eng).
  Commission Implementing Regulation (EU) No 844/2012 of 18 September 2012 setting out the provisions necessary for the implementation of the renewal procedure for active substances, as provided for in Regulation (EC) No 1107/2009 of the European Parliament and of the Council concerning the placing of plant protection products on the market
- Regulation (EC) 1272/2008 (https://eur-lex.europa.eu/legal-content/EN/TXT/PDF/?uri=CELEX:32008R1272).
  Regulation (EC) No. 1272/2008 of the European Parliament and of the Council of 16 December 2008 on classification, labelling and packaging of substances and mixtures, amending and repealing Directives 67/548/EEC and 1999/45/EC, and amending Regulation (EC) No 1907/2006
- Regulation (EU) 2023/574 (https://eur-lex.europa.eu/eli/reg_impl/2023/574/oj/eng).
  Commission Implementing Regulation (EU) 2023/574 of 13 March 2023 setting out detailed rules for the identification of unacceptable co-formulants in plant protection products by Regulation (EC) No 1107/2009 of the European Parliament and of the Council
- Regulation 395/2005 (https://eur-lex.europa.eu/legal-content/EN/ALL/?uri=celex%3A32005R0396).
  Regulation (EC) No 396/2005 of the European Parliament and of the Council of 23 February 2005 on maximum residue levels of pesticides in or on food and feed of plant and animal origin and amending Council Directive 91/414/EEC
- Directive 79/117/EEC (https://eur-lex.europa.eu/eli/dir/1979/117/oj).
  Council Directive 79/117/EEC of 21 December 1978 prohibiting the placing on the market and use of plant protection products containing certain active substances
- Directive 91/414/EEC (https://eur-lex.europa.eu/eli/dir/1991/414/oj/eng).
  Council Directive 91/414/EEC of 15 July 1991 concerning the placing of plant protection products on the market

Annex 3: URLs Relevant for the Vision of EFSA on Endocrine-Disrupting (ED) Compounds, Including Glyphosate

This is a short selection; many other documents, including those provided by action groups, are available.

- Scientific criteria for identification of endocrine disruptors and appropriateness of existing test methods for assessing effects mediated by these substances on human health and the environment. This is the basic document on EDs in the EU, published in 2013.
  https://www.efsa.europa.eu/en/efsajournal/pub/3132
- The document of 2013 was supplemented with a document in 2018: Guidance for the identification of endocrine disruptors in the context of Regulations (EU) No. 528/2012 and (EC) No. 1107/2009
  https://www.efsa.europa.eu/en/efsajournal/pub/5311
- Letter from the Commission to Dr. Url, acting executive director of EFSA
  https://www.efsa.europa.eu/sites/default/files/applications/ECletterNAS.pdf
- Paper by Viviani et al. on the identification of endocrine-disrupting properties
  https://efsa.onlinelibrary.wiley.com/doi/abs/10.2903/sp.efsa.2023.EN-7748

Annex 4: Historical Overview of the main Events Related to the 2023 Glyphosate Dossier in the EU

Date	Reference	Description
Before 1991		Pesticides were registered in each member state separately.
1991	Council Directive 91/414/EEC	Adoption of a 2-way registration procedure: The active substance was reviewed on the EU level, and the use of the formulation was reviewed at the level of the member state. Any pesticide registered before 1991 was due to re-registration in a work program that should finish in 2010.
1995		Review of glyphosate for re-registration by Germany as member state rapporteur (MR).
1/7/2002	Commission Directive 2001/99/EC	Glyphosate was approved in the EU for a period of 10 years; the authorization expired in 2012.
2009		EPA reviews its position on glyphosate, but the conclusion of 1991 remains: Glyphosate does not pose a cancer problem.

(continued)

Date	Reference	Description
21/10/2009	Regulation EC 1107/2009	Rules concerning placing on the market of plant protection products (PPP) replacing older Directives 79/117/EEC and 91/414/EEC. This is, until now, the most important EU Regulation. The execution of Regulation EC 1107/2009 is laid down in Regulation 546/2011.
20/12/2013		Re-registration of glyphosate was ordered, taking the meanwhile adopted Regulation EC 1107/2009 into account. The applicant submitted the necessary documents, and the rapporteur member state (Germany) reviewed the documents and submitted the conclusion to the European Food Safety Authority (further called the Authority, EFSA) and the Commission on 20/12/2013.
20/4/2015	https://doi.org/10.1016/S1470-2045(15)70134-8	IARC published its findings in a short notice in Lancet Oncology.
29/4/2015		The Commission ordered EFSA to review the findings of IARC; the deadline for submission of the review was set to 30/10/2015.
29/7/2015	IARC Monograph 112	IARC publishes the full report on glyphosate (together with four other pesticides) and concludes that glyphosate is "category 2A: probably carcinogenic to humans."
30/10/2015		The draft review of EFSA was submitted to the Commission on 30/10/2015, and this review was sent to the Standing Committee on Plants, Animals, Food and Feed on 28/1/2016 (further called: standing committee or PAFF).
12/11/2015	EFSA journal 13(2015)4302	EFSA explains its position with respect to the carcinogenicity of glyphosate and its view on the review of IARC.
13/4/2016	P8_TA(2016)0119	Proposal to renew the approval of glyphosate for 7 years but the proposal was not adopted by the European Parliament.
29/6/2016		The Commission called for a vote on the renewal proposal but no majority was obtained. The opinion of the Committee for Risk Assessment of the European Chemical Agency (ECHA) will be asked for. The Commission prolongs the approval for using glyphosate for 6 months after the arrival of the opinion of ECHA.
15/3/2017		The Risk Assessment Committee (RAC) of ECHA concludes by consensus that glyphosate does not cause cancer in humans and should not be classified as a substance that causes genetic damage. Similar conclusions are presented by EFSA and the World Health Organization meeting on Pesticide Residues (WHO-JMPR).

(continued)

Date	Reference	Description
16/5/2017		The Commission agreed that the discussions on the renewal of the approval of glyphosate should start with the member states.
15/6/2017		ECHA publishes its conclusion on the carcinogenicity of glyphosate. Hence the deadline for approval is 31/12/2017.
7/9/2017	EFSA Journal 15(2017)4979	EFSA document on "the peer review of pesticide risk assessment of the potential endocrine disrupting properties of glyphosate."
5–6/10/2017		A further round of discussion with the member states. The member states have received the EFSA conclusion on 7 September 2017.
6/10/2017	European Citizen Initiative	The European Commission officially received a successful European Citizens' Initiative with validated signatures of more than one million citizens in at least seven member states.
24–25/10/2017	P8_TA-PROV(2017)0395	The Commission held another round of discussion with the member states. The proposal is to renew the approval for 10 years. Since there is no support for this prolongation, the PAFF proposes to prolong the approval for 5 years.
27/11/2017		A qualified majority in favor of the proposal by the Commission to renew the approval of glyphosate for 5 years was reached by the appeal committee (18 member states in favor, 9 member states against and 1 member state abstained).
14/12/2017	C(2017)8419	The Commission adopted the act to renew the approval of glyphosate for 5 years. The use of glyphosate is limited till 15/12/2022.
6/2/2018	P8_TA-PROV(2018)0022	The Commission decides to set up a special committee on the Union's authorization procedure for pesticides. The Committee is called PEST.
13/9/2018	P8_TA(2018)0356	European Parliament resolution of 13 September 2018 on the implementation of the Plant Protection Products Regulation (EC) No. 1107/2009 (2017/2128(INI)).
17/9/2018	PR\1162183.docx	Draft report of the PEST Committee.
18/12/2018	A8-0475/2018	Final report of the PEST committee: "on the Union's authorization procedure for pesticides (2018/2153 (INI))."
16/1/2019		Vote of the EU Parliament on the report of the PEST committee (526 votes in favor, 66 against, and 72 abstentions).

(continued)

Date	Reference	Description
10/5/2019	Commission Implementing Regulation (EU) 2019/724	Four member states (France, Hungary, The Netherlands, and Sweden), further called the Rapporteur Member States or RMS are appointed to act jointly as rapporteurs for the next assessment of glyphosate. This group of member states will be known as the Assessment Group on Glyphosate (AGG).
15/12/2019	According to Regulation 844/2012	A consortium of eight companies (among them BAYER as the most important), further called the Glyphosate Renewal Group of GRG, submitted an application for the renewal of the approval for glyphosate.
8/6/2020		The GRG submit a supplementary dossier for approval to the four RMSs. The RMS checks the completeness of the dossier according to Article 8 of Regulation 844/2012.
18/8/2020		The AGG members inform the applicant (GRG) and EFSA on the admissibility of the dossier.
15/6/2021		EFSA receives the initial evaluation of the glyphosate dossier in the Renewal Assessment Report (RAR). This is an 11,000-page document made by the member states France, Sweden, Hungary, and the Netherlands.
23/9/2021	According to Art 12 of Regulation 844/2012	EFSA distributes the RAR to the European member states and the GRG for consultation and comments. At the same time, EFSA starts a public consultation initiative of 60 days. 416 submissions were received.
24/11/2021		All information, including comments, is forwarded to the Commission, and the comments are forwarded to the RMS for compilation and inclusion in the dossier. The applicants are requested to respond to the comments.
10/5/2022		EFSA and ECHA request an adaptation of the timeline in view of the massive amount of information.
30/5/2023		ECHA confirms the current hazard classification of glyphosate as a severe eye irritant and as toxic to aquatic life. Classification as carcinogenic is not justified.
30/9/2022		The AGG submits an updated RAR to EFSA. Herein, the additional information gathered through the different consultation rounds is included.
14/10/2022		EU member states had to vote at the Standing Committee on Plants, Animals, Food and Feed on the 1-year extension; no qualified majority was reached.
15/11/2022		The Commission referred to the Appeal Committee for further deliberation. Here again, no sufficient support in favor not against the 1 year prolongation was obtained.

(continued)

Date	Reference	Description
2/12/2022	Adopting Regulation (EU 2022/2364	The EU Commission extends the decision for 1 year till 15/12/2023.
22/12/2022		EFSA publishes the reports from the peer review expert meetings held between 14/11/2022 and 2/12/2022. The conclusions are included.
9/2/2022	According to art 13(3) of Regulation 844/2012	It was decided during a telephone conference between EFSA, RMS, and the Commission that an Expert consultation on mammalian toxicology, residues, environmental fate, and behavior, and ecotoxicology must be organized and that additional information from the applicant is needed.
May 2022		A final consultation on the conclusions arising from the peer review of the RAR was held using a written procedure.
6/7/2023		EFSA informs the GRG, EU member states, and the Commission of their conclusion: EFSA did not identify any critical areas of concern.
26/7/2023		The conclusion of EFSA entitled: "peer review of the pesticide risk assessment of the active substance glyphosate" is published in the EFSA journal https://doi.org/10.2903/j.efsa.2023.8164
25/8/2023		The EU Peer Review Report of the risk assessment of glyphosate is made public. Background documents will be published later.
13/9/2023		EFSA publishes the final RAR including the last batch of background documents on open.efsa
13/10/2023		Vote the prolongation in the Standing Committee of Plants, Animals, Food and Feed (SCOPAFF). The requested majority to renew of reject the project was not reached.
16/11/2023		Vote on the prolongation at the Appeal Committee. No qualified majority on the Commission's proposal to renew the approval of glyphosate for 10 years was not reached.
29/11/2023		The Commission Implementing Regulation (EU) 2023/2660, hereby renewing the approval of the active substance glyphosate.

The renewal of the permit to use glyphosate was published on 28/11/2023. https://eur-lex.europa.eu/legal-content/EN/TXT/PDF/?uri=OJ:L_202302660

COMMISSION IMPLEMENTING REGULATION (EU) 2023/2660 of 28 November 2023

renewing the approval of the active substance glyphosate in accordance with Regulation (EC)

No. 1107/2009 of the European Parliament and of the Council and amending Commission

Implementing Regulation (EU) No. 540/2011

Restrictions Implied in the Approval of Further Use of Glyphosate

- The *prohibition for use as a desiccant* (with the intention to control the time point of harvest or to optimize the threshing).
- The *setting of maximum limits for five impurities in glyphosate* (i.e., in the material manufactured). These limits are set to ensure that humans and the environment are fully protected.
- The *requirement for member states to pay particular attention to specific aspects when carrying out risk assessments* (e.g., the protection of small herbivorous mammals, such as voles, and non-target plants, such as wildflowers) *and to set risk mitigation measures* to ensure that non-target organisms and the environment are protected.
- The *setting of maximum application rates* that shall not be exceeded unless the outcome of the risk assessment undertaken for the specific uses for which authorization is applied demonstrates that a higher rate does not lead to any unacceptable effects on small herbivorous mammals.
- An *obligation for the applicant to submit information on the possible indirect impacts on biodiversity* within 3 years from the availability of an appropriate guidance document. The Commission will request EFSA to develop the necessary guidance.

Annex 5: Files in Renewal Assessment Report According to 1107/2009 and Harmonized Labeling 1272/2008

Files provided by EFSA as Supporting documents for EFSA-Q-2020-00140 (Sequence as shown on the website)

1. Peer review report glyphosate (61.5 MB)
2. Glyphosate AIRV_APPL.zip (4.4 MB)
3. Glyphosate AIRV_USSSD_CP_M.zip (22.6 MB)
4. Glyphosate AIRV_USSSD_CA_M_7-10.zip (113.9 MB)
5. Glyphosate AIRV_USSSD_CA_M_1-6.zip (132.8 MB)
6. Glyphosate AIRV-USSSD_CA_L_1-5.zip (102 MB)
7. Glyphosate AIRV_USSSD_A-I,N...zip (43.7 MB).
8. Glyphosate AIRV_USSSD_CA_L_6-10.zip (106.9 MB)
9. Glyphosate AIRV_USSSD_CP_L.zip (29.9 MB)
10. 2022-03-14 Glyphosate AIRV_Additional information request.pdf (276.9 KB)
11. Glyphosate (AIRV) Reply additional information request GTF.pdf (91.1 KB)
12. Consolidated list of newly available publications after commenting period EFSA+AGG march 2023_public XLSX
13. Glyphosate Final RAR public.zip (428.8 MB)

- AIRV: the fifth review process.
- USSSD: Unified System for the Submission of Scientific Data. This is a standardized system in the EU to be used by applicants in the context of regulatory processes.
- CA: Competent Authority. This dossier is set up by the member state/rapporteur who is evaluating the overall risk assessment for the compound. This includes topics like toxicology, exposure, and environmental issues. Human health is the central theme.
- CP: Consumer Protection. This dossier is compatible with the CA dossier with a focus on consumer health. This dossier is set up by EFSA in collaboration with the member state. Items like maximum residue levels and exposure through food are discussed.

Most of these files are further subdivided.

The *Peer review report glyphosate* consists of six individual files all made public in July 2023. These files are primarily intended for the finetuning of the RAR. In general, many remarks and questions were formulated from the member states or the experts to EFSA or the applicants, and the appropriate answers were given.

1. Comments on the assessment reports (1131 pages). In this file, all comments from the member states and from EFSA on the RAR are summarized.
2. Peer review report on glyphosate (2930 pages). This contains all remarks from EFSA to the individual providers of information. These files are not accessible.
3. Pesticides peer review meeting reports (672 pages). There were five formal telephone meetings with experts (TC 80–84) each with a specific topic: 80: Mammalian toxicology, 81: Environmental fate and behaviour, 82: Ecotoxicology, 83: Residues, and 84: Mammalian toxicology–ecotoxicology joint session.
4. Evaluation tables (1093 pages). This file is congruent with File 2 and is confidential.
5. Comments on the additional information assessment (372 pages). Congruent with Files 2 and 4. This is confidential.
6. Comments on the draft EFSA conclusion (104 pages). This file contains the final remarks on the peer review.

The *Glyphosate AIRV_APPL.zip (4.4 MB)* is a 25-page document with the demographic information on the applicant and on the identification (a.o. CAS nomenclature) of the active compound. This file includes a list of additional studies (ongoing, not reviewed…) marked as "data protected" or "data confidentiality" and will not be easily accessible to the public.

The *Glyphosate AIRV_USSSD_CP_M.zip (22.6 MB)* file contains 12 sections which are discussed in detail in the text.

The *Glyphosate AIRV_USSSD_CA_M_7-10.zip (113.9 MB)* file contains four separate files (7–10) which are discussed in the text.

The *Glyphosate AIRV_USSSD_CA_M_1-6.zip (132.8 MB)* contains six files (1–6) which are discussed in the text.

The *Glyphosate AIRV-USSSD_CA_L_1-5.zip (102 MB)* file contains five sections (1–5) on the identity, physicochemical properties, analytical methods, and an extensive list of toxicity studies and scientific research results mentioned in the renewal dossier (fifth file). Some of the studies are data protected, and some are first submission in the EU (most of the studies have been submitted in the 2015—AIR 2—or the 2002 renewal process).

The *Glyphosate AIRV_USSSD_A-I,N...zip (43.7 MB)* file contains 18 documents (A–N) delivered by the GRG with clarifications on, e.g., ownership, the identity of the molecule, own conclusions, etc.

The *Glyphosate AIRV_USSSD_CA_L_6-10.zip (106.9 MB)* is the continuation of the CA_L_1-5 dossier and contains five separate files (Sections 6–10).

The *Glyphosate AIRV_USSSD_CP_L.zip (29.9 MB)* file contains 12 sections on identity, properties, analytical methods, toxicity, ecotoxicity, etc.

The *2022-03-14 Glyphosate AIRV_Additional information request.pdf (276.9 KB)* is the letter of the EFSA pesticide peer review unit to the GRG with the request for additional information on parts of the dossier.

The *Glyphosate (AIRV) Reply additional information request GTF.pdf (91.1 KB)* is the reply letter of the GRG to the request for additional information. The letter refers to this information but is not included in the letter.

The *Consolidated list of newly available publications after commenting period EFSA+AGG march 2023_public XLSX* is an interesting list with 207 references added to the dossier. Nine references were more than 15 years old, the most recent papers being from 2022.

The *Glyphosate Final RAR public.zip (428.8 MB)* is the final RAR. It contains 30 volumes (and subdivisions) corresponding to the CA and CP-type files mentioned earlier. In addition, a number of Excel files are provided with the summary of the study results in table format.

Taken together, there is general agreement that this is a large dossier with much information and, corresponding to the strategy of EFSA, transparent. Because of data confidentiality or data protection, it is often not easy to locate a particular study or results. The codes are quite complex but, in view of the size of the dossier and the approach by EFSA, comprehensible.

Additional information is available at the following URLs.

Renewal assessment report according to 1107/2009 and harmonized labeling 1272/2008.	Peer review report
VOLUME 1, 868p, 6/2021, initial RAR Volume 1, ED assessment for humans, 444p, 6/2021, initial RAR Volume 2, 6/2021, 283p Volume 3, B.1 (AS), 7p, idem Volume 3, B.2 (AS), 48p Volume 3, B.3 (AS), 27p Volume 3, B.4 (AS), 7p	Part 1: Comments on assessment report Part 2: Reporting tables Part 3: Glyphosate expert meeting Part 4: Evaluation tables Part 5: Comments on additional information Part 6: Comments on draft EFSA conclusion

Title	URL	Pages	
Glyphosate Volume 1	https://www.glyphosate.eu/app/uploads/2021/10/Glyphosate_RAR_01_Volume_1_2021-08-10_sanitised_V2_amended.pdf	868	1. Statement of subject matter and purpose for which this report has been prepared and background information on the application 2. Summary of active substance hazard and of product risk assessment 3. Proposed decision with respect to the application
Glyphosate Volume 1 ED assessment for humans	https://www.glyphosate.eu/app/uploads/2021/10/Glyphosate_RAR_02_Volume_1_2.10.2_ED_2021-08-10_sanitised-V2.pdf	444	2.10. Endocrine disrupting properties
Glyphosate Volume 2	https://www.glyphosate.eu/app/uploads/2021/10/Glyphosate_RAR_03_Volume_2_2021-08-10_sanitized_part1_V2_amended.pdf	283	A. List of the tests, studies, and information submitted
Glyphosate Volume 3—B.1. (AS)	https://www.glyphosate.eu/app/uploads/2021/10/Glyphosate_RAR_04_Volume_3CA_B--1_2021-08-10_sanitized-none_V2.pdf	7	B.1. Identity
Glyphosate Volume 3—B.2. (AS)	https://www.glyphosate.eu/app/uploads/2021/10/Glyphosate_RAR_05_Volume_3CA_B--2_2021-08-10-sanitized_V2.pdf		B.2. Physical and chemical properties of the active substance
Glyphosate Volume 3—B.3. (AS)	https://www.glyphosate.eu/app/uploads/2021/10/Glyphosate_RAR_06_Volume_3CA_B--3_2021-08-10-sanitized_V2.pdf	27	B.3. Data on application
Glyphosate Volume 3—B.4. (AS)	https://www.glyphosate.eu/app/uploads/2021/10/Glyphosate_RAR_07_Volume_3CA_B--4_2021-08-10-sanitized-none_V2.pdf	7	B.4. Further information
Glyphosate Volume 3—B.5. (AS)	https://www.glyphosate.eu/app/uploads/2021/10/Glyphosate_RAR_08_Volume_3CA_B-5_2021-08-10_sanitized_V2_amended.pdf	744	B.5. Methods of analysis

(continued)

Title	URL	Pages	
Glyphosate Volume 3—B.6.1.–B.6.10. (AS)	https://www.glyphosate.eu/app/uploads/2021/10/Glyphosate_RAR_09_Volume_3CA_B-6.1-B-6.2_2021-08-10_sanitized_V2.pdf	595	B.6. Toxicology and metabolism data B.6.1. Absorption, distribution, metabolism and excretion in mammals B.6.2. Acute toxicity B.6.3. Short-term toxicity B.6.4. Genotoxicity B.6.5. Long-term toxicity and carcinogenesis B.6.6. Reproductive toxicity B.6.7. Neurotoxicity B.6.8. Other toxicological studies B.6.9. Medical data and information B.6.10 rReferences
Glyphosate Volume 3—B.6.3. (AS)	https://www.glyphosate.eu/app/uploads/2021/10/Glyphosate_RAR_10_Volume_3CA_B--6.3_2021-08-10_sanitized_V2.pdf	281	B.6. Toxicological and metabolism data
Glyphosate Volume 3—B.6.4. (AS)	https://www.glyphosate.eu/app/uploads/2021/10/Glyphosate_RAR_11_Volume_3CA_B-6.4_2021-08-10_sanitized_V2_amended.pdf	396	B.6.4. Genotoxicity
Glyphosate Volume 3—B.6.5.	https://www.glyphosate.eu/app/uploads/2021/10/Glyphosate_RAR_12_Volume_3CA_B--6.5_2021-08-10_sanitized_V2.pdf	361	B.6.5. Long-term toxicity and carcinogenesis
Glyphosate Volume 3—B.6.6.	https://www.glyphosate.eu/app/uploads/2021/10/Glyphosate_RAR_13_Volume_3CA_B--6.6_2021-08-10_sanitized_V2.pdf	450	B.6.6. Reproductive toxicity
Glyphosate Volume 3—B.6.7.–B.6.10. (AS)	https://www.glyphosate.eu/app/uploads/2021/11/Glyphosate_RAR_14_Volume_3CA_B-6.7-B-6.10_2021-08-10.pdf	1011	B.6.7. Neurotoxicity B.6.8. Other toxicological studies B.6.9. Medical data and information B.6.10. References

(continued)

Title	URL	Pages	
Glyphosate Volume 3—B.7.1.–B.7.4. (AS)	https://www.glyphosate.eu/app/uploads/2021/11/Glyphosate_RAR_15_Volume_3CA_B-7.1-B-7.4_2021-08-10.pdf	804	B.7. Residue data B.7.1. Storage and stability of residues B.7.2. Metabolism, distribution, and expression of residues B.7.3. Magnitude of residue trials in plants B.7.4. Feeding studies
Glyphosate Volume 3—B.7.5.–B.7.8. (AS)	https://www.glyphosate.eu/app/uploads/2021/11/Glyphosate_RAR_16_Volume_3CA_B-7.5-B-7.8_2021-08-10.pdf	404	B.7.5. Effects of processing B.7.6. Residues in succeeding or rotational crops B.7.7. Other studies B.7.8. References
Glyphosate Volume 3—B.8. (AS)	https://www.glyphosate.eu/app/uploads/2021/11/Glyphosate_RAR_17_Volume_3CA_B-8_2021-08-10.pdf	1405	B.8. Environmental fate and behaviour
Glyphosate Volume 3—B.8.5. (AS)	https://www.glyphosate.eu/app/uploads/2021/11/Glyphosate_RAR_18_Volume_3CA_B-8_8.5.-Monitoring-data_2021-08-10.pdf	741	B.8.5. Monitoring data concerning fate and behaviour of the active substance, metabolites, degradation and reaction products
Glyphosate Volume 3—B.9. (AS)	https://www.glyphosate.eu/app/uploads/2021/11/Glyphosate_RAR_19_Volume_3CA_B-9_2021-08-10.pdf	846	B.9. Exotoxicological data
Glyphosate Volume 3—B.9. (AS)	https://www.glyphosate.eu/app/uploads/2021/10/Glyphosate_RAR_20_Volume_3CA_B-9_Appendix-literature-search_2021-08-10_sanitized_V2.pdf	580	B.9. Ecotoxicological data appendix literature data
Glyphosate Volume 3—B.1. (PPP) MON 52276	https://www.glyphosate.eu/app/uploads/2021/10/Glyphosate_RAR_21_Volume_3CP_MON-52276_B-1_2021-08-10_sanitized-none_V2.pdf	6	B.1. Identity
Glyphosate Volume 3—B.2. (PPP) MON 52276	https://www.glyphosate.eu/app/uploads/2021/10/Glyphosate_RAR_22_Volume_3CP_MON-52276_B-2_2021-08-10-sanitized_V2.pdf	26	B.2. Physical and chemical properties of the plant protection product name
Glyphosate Volume 3—B.3. (PPP) MON 52276	https://www.glyphosate.eu/app/uploads/2021/10/Glyphosate_RAR_23_Volume_3CP_MON-52276_B-3_2021-08-10_sanitized_V2.pdf	33	B.3. Data on application and efficacy

(continued)

Title	URL	Pages	
Glyphosate Volume 3—B.4. (PPP) MON 52276	https://www.glyphosate.eu/app/uploads/2021/10/Glyphosate_RAR_24_Volume_3CP_MON-52276_B-4_2021-08-10-sanitized_V2.pdf	16	B.4. Further information
Glyphosate Volume 3—B.5. (PPP) MON 52276	https://www.glyphosate.eu/app/uploads/2021/10/Glyphosate_RAR_25_Volume_3CP_MON-52276_B-5_2021-08-10-sanitized_V2.pdf	47	B.5. Methods of analysis
Glyphosate Volume 3—B.6. (PPP) MON 52276	https://www.glyphosate.eu/app/uploads/2021/10/Glyphosate_RAR_26_Volume_3CP_MON-52276_B-6_2021-08-10_sanitized_V2_amended.pdf	165	B.6. Toxicology and metabolism data and assessment of risks for humans
Glyphosate Volume 3—B.7. (PPP) MON 52276	https://www.glyphosate.eu/app/uploads/2021/10/Glyphosate_RAR_27_Volume_3CP_MON-52276_B-7_2021-08-10-sanitized-none_V2.pdf	4	B.7. Residue data
Glyphosate Volume 3—B.8. (PPP) MON 52276	https://www.glyphosate.eu/app/uploads/2021/10/Glyphosate_RAR_28_Volume_3CP_MON-52276_B-8_2021-08-10-sanitized_V2_amended.pdf	80	B.8. Environmental fate and behaviour
Glyphosate Volume 3—B.9. (PPP) MON 52276	https://www.glyphosate.eu/app/uploads/2021/10/Glyphosate_RAR_29_Volume_3CP_MON-52276_B-9_2021-08-10_sanitized_V2.pdf	471	B.9. Ecotoxicological data and assessment of risks for non-target species
Glyphosate Volume 3—B.9. (PPP) MON 52276	https://www.glyphosate.eu/app/uploads/2021/10/Glyphosate_RAR_30_Volume_3CP_MON-52276_B-9_App_lit_biodiv_2021-08-10_sanitized_V2.pdf	97	Appendix A tot volume 3—(PPP) MON 52276—B.9. Literature data on biodiversity
Glyphosate list of endpoints	https://www.glyphosate.eu/app/uploads/2021/10/Glyphosate_RAR_40_List-of-Endpoints_2021-08-10_sanitized_V2.pdf	169	
Appendix E. ecotoxicity	https://www.glyphosate.eu/app/uploads/2021/10/Glyphosate_RAR_Appendix-E_ecotox_2021-09-15.xlsm		Excel file
Appendix E toxicity	https://www.glyphosate.eu/app/uploads/2021/10/Glyphosate_RAR_Appendix-E_tox_2021-09-15.xlsm		Excel file (animal studies—latest: 2019, oldest: 1979)

(continued)

Annex 6: Information of the Files Released by EFSA in 2018 Called "Public Files"

Identification of the studies

Nr	Title	Identification 1	Identification 2	Identification 3
1	104-week combined chronic feeding/oncogenicity study in rats with 52-week interim kill (results after 104 weeks)	INVERESK research international report 7867	IRI Project No. 438623	CHA Doc. No. 153 GLY
2	Glyphosate acid: 2-year dietary toxicity and oncogenicity study in rats: Part 1	CTL/PR1111/ Regulatory report		
3	Glyphosate acid: 2-year dietary toxicity and oncogenicity study in rats: Part 2	Contains the details of Study 2 (tables, figures, etc., individual pathology data).		
4	24-month oral chronic toxicity and oncogenicity study in rats	HR-001	IET 94-0150	HA 5.5.1/01
5	Chronic study of glyphosate administered in feed to Albino rats	MSL-10495	R.D. No. 1014	Study No. 87122 Project No. ML-87-148
6	Combined chronic toxicity and carcinogenicity study with glyphosate technical in Wistar rats	Study No. toxi: 886.C.C-R	Toxi-886/1996	ES-GPT-C.C-R
7	Glyphosate technical: dietary combined chronic toxicity/ carcinogenicity study in the rat: Part 1	SPL project number 2060-0012		
8	Glyphosate technical: dietary combined chronic toxicity/ carcinogenicity study in the rat: Part 2	Cfr Study 7		
9	Glyphosate 104-week dietary carcinogenicity study in mice	IRI Project No. 438618	IRI Report No. 7793	
10	A chronic feeding study of glyphosate (roundup technical) in mice—revised report (24 months)	BDN-77-420	Project No. 77-2061	
11	Carcinogenicity study with glyphosate technical in Swiss Albino mice; final report	Study No. toxi: 1559. carci-M		
12	HR-001: 18-month oral oncogenicity study in mice	IET 94-0151		

(continued)

Nr	Title	Identification 1	Identification 2	Identification 3
13	Glyphosate technical: dietary carcinogenicity study in the mouse	SPL project 2060-0011		
14	A lifetime feeding study of glyphosate (roundup technical) in rats, Part 1	BDN-416	Project No. 77-2062	
15	A lifetime feeding study of glyphosate (roundup technical) in rats, Part 2 (details of Study 14)		Project No. 77-2062	
16	A lifetime feeding study of glyphosate (roundup technical) in rats, Part 3		Project No. 77-2062	

Commentary 1: For some studies, multiple identification tags are used. Most of them are interns to the organization performing the analysis

Commentary 2: Studies 2 and 3 are one study, further defined as Study 2. The same holds for Studies 7 and 8, further defined as Study 7. Studies 14, 15, and 16 are part of one study, defined as Study 9

Information on the performing lab and the study

Nr	Sponsor	Performing lab/ organization	Number of pages	Important data
1	Cheminova (Denmark)	Inveresk Research International IRI (Scotland)	1510 Volume 1: 1–56: results, 56-438: tables Volume 2: 439–445 and 1123–1349 Volume 3: 1350–1510 individual results.	7/4/1993 (report to the management)
2	Syngenta Limited (UK)	Central Toxicology Laboratory (UK)	2000 1–37: results 38–427: figures, tables, appendices	Start: 1/4/1998 End experiment: 16/10/2000 Report: 15/3/2001
3	Cfr Study 2			
4	Sankyo Co (Japan)	The Institute of Environmental Toxicology—IET (Japan)	2000 8 volumes Volume 1: 14–305: text, figures, and tables Volumes 2–8: appendices	Contract: 8/12/1994, final report: 1/7/1997

(continued)

Nr	Sponsor	Performing lab/ organization	Number of pages	Important data
5	Monsanto Agricultural Company	Monsanto aAgricultural Company (USA)	2175 Volumes 1–6	26/9/1990
6	Feinchemie Schweba (Germany)	Toxicological Department, Rallis Research Centre (India)	615 1–90: results and conclusions 91–615: tables, figures, appendices	20/7/1996
7	Nufarm Asia (Malaysia)	Harlan Laboratories (UK)	2464 Volume 1: 1–762 1–67: results 67–2464: tables, appendices	Start experiment: 7/4/2006 Final quality audit: 23/4/2009
8	Cfr No. 7			
9	Cheminova (Denmark)	Inveresk Research International (IRI) Scotland	750 27–33: results 35–750: figures, tables, appendices	Start of the experiment: 30/10/1989, end of the experiment: 23/12/1990. Final report: 7/4/1993
10	Monsanto (USA)	Bio/Dynamics Inc (BDN)	Volume 1: T 1-1–T 20C-1 results Volumes 2–8: figures, tables, appendices	Start of the treatment: 31/3/1980 Original report: 21/7/1983 Revised report: …/3/1986
11	Feinchemie Schwebda (Germany	Toxicology Department Rallis Research Centre (India)	583 1–109: results 110–582: tables, figures, appendices	(start at 23/12/1997, end at 29/6/1999) Report delivery: 10/10/2001
12	Sankyo Co (Japan)	The Institute of Environmental Toxicology (Japan) Mitsukaido Laboratories, Ibaraki (Japan)	1244 Volume 1: 1–64: results; 65–184: tables, figures Volumes 2–8: appendices	Start: 13/12/1994 (contract)–18/6/1997 (final report)
13	Nufarm Asia (Malaysia)	Harlan Laboratories (UK)	1220 Volume 1: 1–48: results 49–1220: tables, figures, appendices	Contract: 4/8/2005 Start of the experiment: 10/10/2005 End of the experiment: 19/11/2007 Final report: 22/4/2009 Report delivery: 22/4/2009

(continued)

Nr	Sponsor	Performing lab/ organization	Number of pages	Important data
14	Monsanto	Bio/dynamics Inc	Volume 1: 1–13: results A1-L1: appendices Volumes II–V	Treatment start: 12/7/1978 Treatment end: 25/8/1980 Report: 18/9/1981
15	Cfr Study 14			
16	Cfr Study 14			

Commentary: The number of pages is the figure mentioned in the document. The real number of pages is usually less because volumes often do not follow a logical linearly increasing numbering

Active compound, guidelines, dose, animals

Nr	Product	Guidelines	Dose	Animals
1	Technical glyphosate, 98.9% purity	EPA pesticide assessment guidelines subdivision F, 83-5 GLP according to EPA (40 CFR part 160) OECD and Japanese MAFF	0, 10, 100, 300, and 1000 mg/kg/day (weekly dose adjustment for 13 weeks)	85 Sprague-Dawley rats/sex/ dose; 50 rats for the oncogenicity study and 35 rats for the toxicity study
2	Glyphosate acid technical 97.6% purity	(p14) OECD guideline 435 (1981), Annex V to Directive 67/548/ EEC, Directive 87/302/EEC, OJEC, L133 (1988, EPA guidelines, OPPTS 870.4300 (1988) and control law of Japan (MITI handbook, 1992)	0, 2000, 6000, and 20,000 ppm	52 male and 52 female Wistar-derived rats
3				
4	HR-001 (the word "glyphosate" is not mentioned). The solubility is 12/g in water which corresponds to the solubility of glyphosate acid	GLP of MAFF in Japan (1984); EPA (FIFRA, 1989) and OECD (1981)	0, 3000, 10,000, and 30,000 ppm corresponding to (for males) 104,354 and 1127 mg/kg/day and (for females) to 114, 393, and 1247 mg/kg/day in the treated groups	50 SPF Sprague-Dawley rats/sex/dose in the main group (oncogenicity study); 30 rats/ sex/dose in the satellite group (toxicity study)
5	Glyphosate technical	EPA FIFRA guideline, Guideline 158.34	0, 2000, 8000, and 20,000 ppm for 24 months	60 Sprague-Dawley rats/sex/ dose

(continued)

Nr	Product	Guidelines	Dose	Animals
6	Glyphosate technical 96% purity	OECD (1982)	0, 100, 1000, and 10,000 ppm	50 Wistar rats/sex/dose 10 male and 10 female rats (controls) and 20 male and 20 female rats (highest dose) for interim killing and search for non-neoplastic changes at 52 weeks
7	Glyphosate technical, 95.7% pure	OECD guidelines No. 453 (1997—ENV/MC/CHEM(98)17), Japanese Ministry of Agriculture (JMAFF, 2005, EPA test guidelines and European Directives 2004/9/EC and 2004/10/EC	1500, 5000, and 15,000 ppm corresponding to 95.0, 316.9, and 1229.7 mg/kg/dag	Three main groups of 51 Wistar rats/sex/dose for 2 years and three satellite groups of 45 Wistar rats/sex/dose for 52 weeks one main control group of 51 Wistar rats/sex/dose and one satellite control group of 12 Wistar rats/sex/dose
8	Cfr Study 7			
9	Technical glyphosate, 97.05% pure	US EPA Pesticide assessment guidelines, F, 83-2	0, 100, 300, and 1000 mg/kg/day	50 CD-1 mice/sex/dose level
10	Glyphosate (roundup technical) 99.7% active compound		0, 1000, 5000, and 30,000 ppm	50 CD-1 mice/sex/dose
11	Glyphosate technical	OECD Guideline No. 451 (1981)	0, 100, 1000, and 10,000 ppm administered for 18 months corresponding to (approximately) 0, 15, 150, and 1500 mg/kg/day	50 MF1 strain Swiss Albino mice/sex/dose (originally obtained from Harlan)

(continued)

Nr	Product	Guidelines	Dose	Animals
12	Possibly a material error. The active compound is mentioned to correspond to chromophenozide which is completely different from glyphosate. The solubility (12 g/L) is suggestive of glyphosate.	MAFF (Japan—1984), EPA-FIFRA (USSA—1989), OECD (1981)	0, 1600, 8000, and 40,000 ppm during 78 weeks (18 months) corresponding to (approximately) 1600 ppm = 150 mg/kg/day, 8000 ppm = 840 mg/kg/day, and 40,000 ppm = 4200 mg/kg/day	50 SPF CD-1 mice/sex/dose
13	Glyphosate technical 95.7% purity	UK GLP standards (1999–2004), OECD No. 451 GLP (1981), JMAFF guidelines, directives 2004/9/EC and 2004/10/EV Japanese JMAFF guidelines US OPPTS 870.4200 carcinogenicity guidelines (1996)	0, 500, 1500, and 5000 ppm equivalent to 0 and approximately 84.7, 266.8, and 945.6 mg/kg/day	51 CD-1 mice sex/dose
14	Glyphosate (roundup technical) 98.7% active compound		0, 30, 100, 300 ppm corresponding to (approximately) 3.4, 11.2, and 34.0 mg/kg/day	50 Sprague-Dawley rats/sex/dose
15	Cfr Study 14			
16	Cfr Study 14			

In the next table is a search for the corresponding entries of these studies. Three groups of data are compared:

1. The data delivered by the GRG: nine rat studies, seven mouse studies (https://www.glyphosate.eu/app/uploads/2021/10/Glyphosate_RAR_12_Volume_3CA_B-6.5_20210-08-10_sanitized.pdf).
2. The analysis as found in the RAR: (https://www.glyphosate.eu/app/uploads/2021/10/Glyphosate_RAR_12_Volume_3CA_B-6.5_2021-08-10_sanitized_V2.pdf).
3. The files made public: nine rat studies and five mouse studies (see above for the administrative information).

The question is whether the files made public are also used for the final judgment of glyphosate. In the table is the correspondence (Y/N) depicted.

Rat Carcinogenicity Studies

RAR				GRG		Public	
Code RAR	Date	Identification	Page	Code GRG	Page	Y/N	Correspondence
CA 5.5/001–/052	2009	2060-0012	1356	B.6.5.1	5	Y	Study 7
CA 5.5/002–/053	2001	/PR1111	1363	B.6.5.2	13	N	
CA 5.5/003	1997	1231[a]	1375	B.6.5.3	24	N	
CA 5.5/004–/053	1997	HR-001 94-0150	1384	B.6.5.4	33	Y	Study 4
CA 5.5/005	1996	886.C.C.-R[a]	1413	B.6.5.5	51	Y	Study 6
CA 5.5/006	1996	/P/5143[a]	1446	B.6.5.6	61	N	
CA 5.5/007, /008, 009	1993	7867 153-GLY	1472	B.6.5.7	75	Y	Study 1
CA 5.5/010–/056	1990	-10495 M-651388-12-1	1480	B.6.5.8	85	N	
CA 5.5/011–/057	1981	77-2062 M-645914-01-1	1500	B.6.5.9	98	Y	Studies 14, 15, 16

[a]Marked as "not reported"

Mouse Carcinogenicity Studies

RAR				GRG		Public	
Code RAR	Date	Identification	Page	Code GRG	Page	Y/N	Correspondence
CA 5.5/012, /013, /014, /015	2009	2060-0011	1510	B.6.5.10	104	Y	Study 13
CA 5.5/016–/058, 059	2001	Toxi:1559. CARCI.M	1515	B.6.5.11	110	Y	Study 11
CA 5.5/017	2017	11921 90017583	1525	B.6.5.12	122	Recalculation of study/016	
CA 5.5/017, /018	1997	94-0151	1531	B.6.5.13	138	Y	Study 12
CA 5.5/020, /021	1989	7793 154-GLY	1543	B.6.5.14	149	Y	Study 9
CA 5.5/022	1988	[a]	1548	B.6.5.15	154	Not reported to AGG	

(continued)

RAR				GRG		Public	
Code RAR	Date	Identification	Page	Code GRG	Page	Y/N	Correspondence
CA 5.5/023–/061	1983	77-2061 M-646425-01-1	1557[b]	B.6.5.16	159	Y	Study 10
CA 5.5/024	1982, revision in 1992	8010[a]	1571	B6.5.17	169	N	

[a]these data are not present in the AGG files
[b]Possible errors in Tables 5.5—132, 134, 135; reference is on rats but should be to mice

Annex 7: Question and Answer on the Methodology for Re-reviewing the Review on the Prolongation of Using Glyphosate

My Mail to EFSA

Aan:Eric.Thevenard@ec.europa.eu	
Norbert Fraeyman	Do 28-11-2024 15:41

Dr. Thévenard,

I was directed to you or your unit with the following.

My question relates to the glyphosate dossier. As I understand it, ECHA and EFSA have studied the documents delivered by the Glyphosate Renewal Group and concluded that glyphosate has no carcinogenic and no endocrine-disrupting properties. Other aspects such as the effect on the microbiome or the impact on pollinators have not been taken up despite evidence for concern because there are no formal normative rules to reflect upon. EFSA has communicated their advice to the EU Commission that proposed to prolong the permit for further use. This proposal was put to a vote on several occasions, but no consensus was reached, so, the decision was left to the commission which decided in favour of prolongation. The use of glyphosate was permitted for 10 years with some restrictions. One point in the decision is that when new evidence is available, the permit has to be revised. My question relates to this last part.

As I read it, the Commission will decide whether the permit has to be reviewed based on new evidence. Is this correct? Who will advise the Commission on this matter? Who will judge whether new evidence is worth considering? Is it EFSA? This would be very doubtful because in this case, EFSA would be judge and party at the same time.

Looking forward to your thoughts about this matter.

Kind regards

Norbert Fraeyman

Ref. Ares(2024)8694859 - 05/12/2024

EUROPEAN COMMISSION
DIRECTORATE-GENERAL FOR HEALTH AND FOOD SAFETY

Food Safety, Sustainability, and Innovation
Pesticides and Biocides

Brussels
SANTE.E.4/MW/gb(2024)9991825

Dear Mr Fraeyman,

Subject: Your email on glyphosate dated 28 November 2024

Thank you for your email (registered under our reference ARES(2024)8510509) sent to me on 28 November 2024, in which you request clarification on how the approval of glyphosate may be reviewed in case new information would come to light.

I understand that you refer to recital 30 of Commission Implementing Regulation (EU) 2023/2660 ([1]) which states that *"Furthermore, the approval of the active substance may be reviewed at any time pursuant to Article 21 of Regulation (EC) No 1107/2009 ([2])."*

Article 21 of Regulation (EC) No 1107/2009 permits the Commission to review the approval of an active substance at any time. The Commission shall take into account the request of Member States to initiate a review. The European Food Safety Authority (EFSA) does not decide whether to initiate a review. The Commission must assess on a case-by-case basis whether a review is justified, or not.

Article 21(1) sets out the conditions for when a review is initiated. In particular, a review may be initiated where new scientific and technical knowledge comes to light that indicates that the substance no longer satisfies the approval criteria laid down in Article 4 of that Regulation.

Article 21(2) permits the Commission to ask the Member States and EFSA for an opinion, or for scientific or technical assistance. However, as for any decision on the approval of an active substance it is the Commission, in its role as risk manager, that decides whether the approval criteria laid down in Regulation (EC) No 1107/2009 are fulfilled and the necessary conditions and restrictions to apply.

([1]) Commission Implementing Regulation (EU) 2023/2660 of 28 November 2023 renewing the approval of the active substance glyphosate in accordance with Regulation (EC) No 1107/2009 of the European Parliament and of the Council and amending Commission Implementing Regulation (EU) No 540/2011. http://data.europa.eu/eli/reg_impl/2023/2660/oj

([2]) Regulation (EC) No 1107/2009 of the European Parliament and of the Council of 21 October 2009 concerning the placing of plant protection products on the market and repealing Council Directives 79/117/EEC and 91/414/EEC. https://eur-lex.europa.eu/legal-content/EN/TXT/?uri=CELEX%3A02009R1107-20210327.

Mr Norbert Fraeyman
Email: Norbert.Fraeyman@UGent.be

Commission européenne/Europese Commissie, 1049 Bruxelles/Brussel, BELGIQUE/BELGIË – Tel. +32 22991111

Indeed, Article 21(3) permits the Commission to amend or withdraw the approval of an active substance if it concludes that the approval criteria provided for in Article 4 are no longer satisfied.

A number of reviews under Article 21 have been carried out leading to either the amendment (e.g. the neonicotinoid active substances - clothianidin, thiamethoxam and imidacloprid ([3]) – now no longer approved) or withdrawal of an approval of an active substance (e.g. ipconazole ([4])).

Yours sincerely,

Electronically signed

Eric THEVENARD
Head of Unit

([3]) Commission Implementing Regulation (EU) No 485/2013 of 24 May 2013 amending Implementing Regulation (EU) No 540/2011, as regards the conditions of approval of the active substances clothianidin, thiamethoxam and imidacloprid, and prohibiting the use and sale of seeds treated with plant protection products containing those active substances. https://eur-lex.europa.eu/eli/reg_impl/2013/485/oj

([4]) Commission Implementing Regulation (EU) 2023/939 of 10 May 2023 withdrawing the approval of the active substance ipconazole in accordance with Regulation (EC) No 1107/2009 of the European Parliament and of the Council, amending Commission Implementing Regulation (EU) No 540/2011 and repealing Commission Implementing Regulation (EU) No 571/2014 http://data.europa.eu/eli/reg_impl/2023/939/oj

Annex 8: Questions and Answers from EFSA on the Accessibility of the 2-Year Carcinogenicity Studies

My Original Questions Are Mixed with the Answer of EFSA

Legal Affairs Services
Parma, 13 February 2025
Ref. LV/PU/mm (2025)—out-33626832

Norbert Fraeyman
University of Ghent

E-mail: norbert.fraeyman@ugent.be

RE: Your application for public access to documents (PAD) of 20 December 2024
Our ref.: PAD 2024/211 (00014421)

Dear Mr. Fraeyman,

I refer to your application for public access to documents (hereinafter, "PAD application") of 20 December 2024, by means of which you requested access to:

> "the animal studies underlying the EFSA decision in 2023 that glyphosate has no effect on human health? Website? Link?."

Additionally, you also requested the following information:

> "4 MEP's requested access to the glyphosate animal studies on 20/3/2016 which was refused by EFSA but this was annulled by the General Court of Justice in March 2019 and 16 animal dossiers were made accessible. Why 16 studies? What were the criteria for this selection? Are these 16 files part of the total number of animal studies that EFSA considered when making their conclusions on glyphosate?" "In addition, 53 files were later made public and here again: are these 53 files part of the EFSA decision process? These files were scrutinized by Prof. Knasmuller (Vienna). What is your reaction to his analysis?"

Your application for public access to documents has been processed in accordance with the provisions of Regulation (EC) No. 1049/2001 on public access to documents (hereinafter, "the PAD Regulation").

Your requests for information have also been processed in accordance with the EFSA Code of Good Administrative Behaviour and the European Code of Good Administrative Behaviour.

By means of this letter, we would like to provide you a reply as follows.

1. Related Background Documents

    - The report from the Pesticides Peer Review experts' discussions on mammalian toxicology and impact on human/animal health, which is available in the Peer Review Report, more precisely refer to the Report of the Pesticides Peer Review TC 80 in Part 3_Peer Review Report_Glyphosate_expert meeting reports_public.pdf (electronic pages 42-373 of 672). The shorter, high-level reports from the Pesticides Peer Review expert meetings are also made available here, providing a record of the points discussed and the brief conclusions reached for each point.

- The final Renewal Assessment Report (final RAR) (2023) and the Peer Review Report, which are publicly available on OpenEFSA, under the "Supporting documents" section at EFSA question number: 2020-00140. More specifically, the information related to mammalian toxicology and its impact on human health can be found in the Final RAR Volume 1 (Chapter 2.6: "Effects on Human and Animal Health" and Chapter 2.10: "Endocrine Disrupting Properties"), in Volumes 3 CA B6 (cf. Volumes 9–14f of the final RAR), Volume 3 CP_MON52276_B6 (cf. Volume 26 of the final RAR).
- Specifically, the evaluation and summary of the individual studies together with their results and conclusions are reported in the abovementioned Volume 3 documents. The reference list containing all the individual studies is listed on pages 173–297 of Volume 3 CA B6 (cf volume 14f) and on pages 97–99 of Volume 3 CP_MON52276_B6 (Volume 26). Animal studies are marked with "Y" in the column "vertebrate study Y/N."

Should you need additional information, or should you be in need of guidance as to where to find public information, you can submit a request for information at any time via Connect.EFSA Portal, selecting "Ask a Question," filling in the form, and choosing "Question about Scientific Application Process" or "General question about EFSA" in the area field.

2. Requests for Information—Replies to Your Questions

Please find a reply to your questions below.

(a) First Request for Information

By means of your first request for information you asked:

> "4 MEP's requested access to the glyphosate animal studies on 20/3/2016 which was refused by EFSA but this was annulled by the General Court of Justice in March 2019 and 16 animal dossiers were made accessible. Why 16 studies? What were the criteria for this selection? Are these 16 files part of the total number of animal studies that EFSA considered when making their conclusions on glyphosate?"

We are pleased to provide you with a reply as follows.

(i) "Why 16 studies?"

As outlined in paragraphs 43–44 of the ruling of the General Court in Case T-329/17 Hautala et al. v EFSA, the action for annulment against EFSA's decision to provide partial access to parts of 75 unpublished study reports on glyphosate, concerned only the 12 "most crucial" studies for the peer review and its conclusion that glyphosate is unlikely to pose a carcinogenic hazard to humans.

During the court proceedings, the applicants confirmed that they sought disclosure only of the parts "material, experimental conditions and methods" and "results and discussion" of the requested studies.

(ii) "What were the criteria for selection?"

The selection of the 12 studies was based on the choice of the MEPs requesting access to the study reports on glyphosate. The criteria for this selection were not communicated to EFSA.

(iii) "Are these 16 files part of the total number of animal studies that EFSA considered when making their conclusions on glyphosate?"

The 12 study reports concerned by the ruling were part of the public access to documents request of the MEPs which had a particular scope that was clarified and reduced in view of their large volume as explained in paragraphs 13–19 of the ruling of the General Court in Case T-329/17 Hautala et al. v EFSA. The PAD applications after clarification concerned:

"all the studies used by EFSA to assess the carcinogenicity of glyphosate and its representative formulation in their entirety, namely:

- the 14 studies on rodents included in the joint submission dossier, including the raw data;
- the historical control data used in those studies, in particular the historical control incidences for renal tumours and for haemangiosarcoma from the laboratories that performed the mouse carcinogenicity studies cited by EFSA;
- all studies that were assessed by EFSA concerning mechanisms of carcinogenicity such as genotoxicity and oxidative stress in relation to glyphosate and its representative formulation in their entirety."

(b) Second Request for Information

With regard to your second request for information you asked:

"In addition, 53 files were later made public and here again: are these 53 files part of the EFSA decision process? These files were scrutinized by Prof. Knasmuller (Vienna). What is your reaction to his analysis?"

We are pleased to provide you with a reply as follows.

The studies cited in the document referred to studies submitted during the previous peer review assessment (EFSA, 2015) and were considered also during the last renewal peer review. Indeed EFSA set a data requirement point for the applicant to consider among others the information provided in the analysis by Armen Nersesyan and Siegfried Knasmueller, submitted during the public consultation (see data requirement 2(2) in Part 4_Peer Review Report_Glyphosate_evaluation tables_public: electronic page 984–985 of 1093, available in the Peer Review Report in Open EFSA, Supporting documents section under EFSA-Q-2020-00140). An evaluation of the concerns raised and their impact on the reliability of the genotoxicity studies was also provided in the RAR.

Please refer to the RMS's assessment in Volume 3 B.6.4.4.44 for a detailed evaluation (page 612 onward). EFSA supported the RMS's assessment. The outcome of the assessment of the genotoxic potential reached in the EFSA peer review was based on a weight-of-evidence approach, allowing a conclusion that glyphosate is unlikely to be genotoxic. For full details, please refer also to the expert consultation points 2.1, 2.2, 2.3, 2.4 (identified following comments by public) and experts' consultation 2.17 at the

Pesticides Peer Review Experts' TC 80 (available in the Peer Review Report in Open EFSA, Supporting documents section under EFSA-Q-2020-00140, Part 3_Peer Review Report_Glyphosate_expert meeting reports). During the experts' consultation on genotoxicity, data gaps were not identified for conducting new genotoxicity studies. Overall, this outcome is in line with the ECHA RAC assessment (page 35 onwards) and EFSA agrees with the conclusions reached in the ECHA RAC Opinion.

3. Confirmatory Application

   To exercise your right to appeal against this decision by a confirmatory application, you may write to EFSA as detailed below. You have 15 working days from receipt of this letter to appeal. In case you submit a confirmatory application, EFSA will inform you of the outcome of this reexamination of your request within 15 working days of receipt, either by granting you access to the documents/protected information or by confirming the refusal/masking. In the latter case, you will also be informed of any further appeal routes available.

Confirmatory applications must be sent:

- By introducing a new application in Connect.EFSA Portal, mentioning the reference to your initial application in relation to which you would like an administrative review of the initial decision, or
- Via e-mail to:

EFSA
Dirk Detken Head of Legal and Affairs Services
Via Carlo Magno 1/A
IT—43126 Parma
Italy
E-mail: dirk.detken@efsa.europa.eu

Yours sincerely,
Luisa Venier
Team Leader
EFSA Public Access to Documents
CC.: M Tiramani (EFSA)

Via Carlo Magno 1A
43126 Parma—ITALY
+39 0521 036 111
www.efsa.europa.eu
Legal Affairs Services
Parma, 13 February 2025
Ref. LV/PU/mm (2025)—out-33626832
Norbert Fraeyman
University of Ghent
E-mail: norbert.fraeyman@ugent.be

RE: Your application for public access to documents (PAD) of 20 December 2024
Our ref.: PAD 2024/211 (00014421)

Dear Mr. Fraeyman,

I refer to your application for public access to documents (hereinafter, "PAD application") of 20 December 2024, by means of which you requested access to:

> "the animal studies underlying the EFSA decision in 2023 that glyphosate has no effect on human health? Website? Link?"

Additionally, you also requested the following information:

> "4 MEP's requested access to the glyphosate animal studies on 20/3/2016 which was refused by EFSA but this was annulled by the General Court of Justice in March 2019 and 16 animal dossiers were made accessible. Why 16 studies? What were the criteria for this selection? Are these 16 files part of the total number of animal studies that EFSA considered when making their conclusions on glyphosate?" "In addition, 53 files were later made public and here again: are these 53 files part of the EFSA decision process? These files were scrutinized by Prof. Knasmuller (Vienna). What is your reaction to his analysis?"

Your application for public access to documents has been processed in accordance with the provisions of Regulation (EC) No. 1049/2001 on public access to documents (hereinafter, "the PAD Regulation").

Your requests for information have also been processed in accordance with the EFSA Code of Good Administrative Behaviour and the European Code of Good.

Final Summary, Conclusion, Critics

The reader is offered an overview of the most important aspects related to the glyphosate dossier in eleven chapters. This dossier was submitted by the Glyphosate Renewal Group (GRG) headed by Bayer, which took over Monsanto in 2018, to the European authorities EFSA and ECHA. Four countries (the Netherlands, Sweden, France, and Hungary) were appointed to review the dossier together with the EFSA, and the summary of results was published in the EFSA journal in 2023. The Commission decided that glyphosate was safe for human use, and permission was granted for 10 years, till 2033. This period of reviewing between 2019/2020 and 2023 was disturbing and marked by controversies, discussions, debates, and overt allegations of malpractice, to name a few. Hence, the decision of the Commission was not applauded by everyone.

The basis for this book was laid in 2019–2020, i.e., when the Belgian Superior Health Council (SHC) published its advice on the further use of glyphosate, and I had the privilege of being a co-author. The Council considered this advice necessary because IARC classified glyphosate as "probably carcinogenic to humans" in 2015. The SHC advised a gradual, controlled phase-out of glyphosate use during a sufficient length period, allowing agriculture to adapt to a glyphosate-free practice.

This book uses elements from the SHC's advice, supplemented with new information originating between 2020 and 2023. The point of view of the SHC has been strengthened in the literature of the last years, justifying the writing of this book. The point of view explained in this book is independent of the SHC and of the organizations I am involved in: the Medical School of the University of Ghent and the University Hospital in Ghent.

EFSA, ECHA, and the Assessment Glyphosate Group were drowned by a flood of documents with results of animal and in vitro experiments, scientific papers, and opinions for a total of 180,000 pages, although this figure depends on who you ask. The 1-year extra above the legal 3 years needed for judging a dossier was acceptable, although most of the results presented in the application were old files up to

N. Fraeyman, *Glyphosate 2023–2033*,
https://doi.org/10.1007/978-3-031-97540-0

40 years, and these files were scrutinized before at least once, most of them twice. Hence, among these pages, only a fraction that had to be reviewed was new.

Any conclusion everyone agrees upon is beyond reach. The interpretations mentioned in this book are as nuanced as possible, but, similar to many attempts before, this will probably not be the ultimate book on glyphosate. Having said this, this is what emerges.

There are sufficient experimental results to consider glyphosate as a carcinogen, but as a carcinogen, it is not comparable to the classics, smoking, and asbestos. Whenever the label "carcinogen" is used, there is a tendency to make this mental reference to the ultimate carcinogens, but this is intellectually not correct. Nevertheless, the "modest" carcinogenicity of glyphosate combined with the abundant use of the pesticide makes it an undeniable risk for carcinogenicity (see Chap. 2 on risk, hazard, and exposure), and this is the main point. Second, it is a personal decision to expose yourself to carcinogens such as smoking and eating red or processed meat, while, similar to children exposed to smoking parents, everybody is unwillingly exposed to glyphosate, even the unconditional advocates and including the producers of the chemical. Notably, this is also the case for micro- and nanoplastics, BPA, and others.

The experimental results on the toxicity for the endocrine system, the nervous system, and the microbiome are solid, trustworthy, and conclusive: Glyphosate is a toxic molecule. Admittedly, not all details are known, but research is ongoing. The toxicity of glyphosate in the environment is beyond any doubt and is generally accepted. Notably, the toxicity of glyphosate for the honey bee, investigated in real-life outdoor conditions with low doses of glyphosate, should be a worry. If we lose (part of) the pollinators, either as a direct effect of glyphosate or as an indirect effect via other mechanisms, or even if glyphosate is only part of the problem, society is in trouble.

The overall conclusion is that glyphosate does not have a simple toxic profile with one target but is a compound with a mixed toxic profile with (possibly) multiple targets. Attention has been fixed on carcinogenicity since the IARC declaration of 2015, but it is not only carcinogenicity that is important. The main obstacle is that up until now, scientific evidence is considered insufficient for the EU advising organizations (EFSA and ECHA) to suggest protective actions more than restrictions on its use. The general feeling is that whenever the precautionary principle could have been applied, it was neglected.

This multiplicity of the toxic phenotype is confusing, and the search for a common denominator is ongoing. This has three important consequences: (1) the absence of indisputable evidence for a common target that would explain the overall toxicity of glyphosate is wrongly interpreted as evidence for the absence of toxicity which fuels the drive and the conviction of advocates of further use of the herbicide; (2) during this scientific research and inevitably linked to research, conflicting opinions have been and will be presented leading ultimately to consensus. However, this takes time which is erroneously interpreted as lack of health-related problems; (3) the likelihood of glyphosate having highly diverse targets is complexed by the fact that glyphosate is a component of the chemical soup we are all exposed to although glyphosate is probably one of the main components. These consequences are excellent grounds for applying the precautionary principle, but this has not happened.

The attitude in this book is not very aggressive. Instead of an request for an instant, radical ban on glyphosate, a gradual downsizing of its use is proposed over a sufficient period to allow adaptation of agriculture practices, and this is exactly what the EU has proposed: a decrease of 50% of pesticide use till 2030 within the frame of the Green Deal. There is, however, a major drawback. The EU proposal is neither a Directive, not a Regulation but a roadmap, allowing the EU and the individual member states to decide how to comply with the goals. This is fine in theory, but we will have to wait for the practical implementation.

Personal Epilogue

This book is not written out of frustration, and I do understand that when making decisions, even solid and evidence-based arguments of scientists cannot be followed entirely. The opposite is true: this book is written from a genuine anxiousness of what could happen—sorry, of what WILL happen—within 10–30 years. At large, all signs suggest an increasing and seemingly irreversible chemical pollution of the environment and all living creatures, and glyphosate is an important component. Not all details are known, but this insecurity is grounds to imply the precautionary principle. We should worry about our children and grandchildren, and this is not an emotional or romantic one-liner; it is real and rational.

Instead of frustration, disappointment prevails. EFSA (and/or the EU Commission) had the opportunity to curb the ever-rising use of glyphosate downwards and to imply their own F2F strategy within the European Green Deal for a 50% reduction of pesticide use by 2030. This reduction is now based on the voluntary actions of the member states, there is no obligation, hence, realization is doubtful.

My impression of the 10-year prolongation of glyphosate use is that the balance between economic arguments and human health is largely skewed to economics, more than on biomedical arguments; human health was secondary to profit. Admittedly, all legal obligations were followed concerning maximum levels of exposure, allowable contamination of food and drinks, restrictions on the use of glyphosate, etc. The legal umbrella was a useful and comforting cover.

Of concern is the attitude of EFSA and the Commission vis-à-vis the need to review their decision of December 2023 based on new information. There is nowadays clear evidence of the biomedical devastating consequences related to glyphosate exposure at concentrations or doses within legally admissible levels of exposure: Glyphosate is a (weak) carcinogen, it has endocrine-disrupting properties (inhibition of aromatase), it disturbs the balance of microorganisms in the gut microbiome and it disturbs normal behavior of the honey bee. It is unclear what additional evidence is required to consider a review process. If this re-reviewing methodology is

N. Fraeyman, *Glyphosate 2023–2033*,
https://doi.org/10.1007/978-3-031-97540-0

not implemented carefully, I predict with a probability adjacent to the certainty that from 2030 to 2033, we will have similar discussions as in 2019–2022/2023 since the experimental results that are available now will be used again—remember the glyphosate story of 2002 and 2012.

Despite all this, there is hope. First, the resistance of herbs to glyphosate will force agriculture to reconsider its use. The resistance is irreversible and will increase since the use of glyphosate will increase—remember the use of antibiotics. Adding more toxic molecules in a mixture to combat herbs is not acceptable; hence, there will be an economic reason to rethink the situation, and this argument might be more decisive compared to any health-related consideration. Second, although alternative agriculture methods without chemicals are sufficient for small-scale practices, they are not suitable for large-scale production needed to satisfy the consumer's wishes. Whether these wishes are real needs or artificially induced by economically interested parties is not discussed here. Third, the Commission decided on the pure compound (despite the use of a formulation in the dossier), but the individual member states will decide on the formulations. Taking the difficulties in 2023 of voting at the EU level into account, the use of glyphosate in some member states might become more restricted than it is now or even banned.

Finally, the efforts of the EU to promote the use of alternatives to glyphosate and to stimulate the search for valid treatment schedules for pesticides in general and glyphosate in particular are appreciated. Some more legal pressure would help.

I rest my case.

The manufacturer's authorised representative in the EU is Springer
Nature Customer Service Centre GmbH, Europaplatz 3, 69115 Heidelberg,
Germany. If you have any concerns regarding our products, please
contact ProductSafety@springernature.com

Printed and bound by CPI Group (UK) Ltd, Croydon, CR0 4YY
30/09/2026
02225590-0002